Why Evolution is Not True

New and Abridged Edition of

"Is Evolution True? Why Darwin's Rejection of Intelligent Design No Longer Makes Sense. Why [it is more and more obvious that] Life Really is a Miracle in the True Sense"

A Ω

Dorrie

SOL OMNIBUS LUCET

Dedicated to all who keep their minds open,
in the *scientific, philosophical* and *spiritual* quests
to discover what may really be true.

With grateful acknowledgement to all the writers and publishers whose work is quoted from or referenced here: John Lennox, Paul Davies, Hugh Ross, Sir Roger Penrose, James Shapiro, Matti Leisola, Esko Valtaoja, William Dembski, Jonathan Witt, Michael Denton, Michael Behe, Mark Eastman, Chuck Missler, Stephen Meyer, Antony Latham, Jonathan Wells, Chris Carter, Art Battson, Charles Darwin, Richard Dawkins, Michael Flannery, John Davidson, Luther Sunderland, Peter Tompkins, James Perloff, Dean Radin, Rupert Sheldrake, P. M. Atwater, Lyn Margulis, Michael Polanyi, Colin Patterson, V. J. Torley, James Tour, Neal Grossman, Sandie Gustus, Steve Taylor, Michael Egnor, Casey Luskin, Tom Bethell, John Morrison, Robert Shedinger, the Journal of the Discovery Institute, *Evolution News*, and many others, too numerous to mention, who have conducted powerful, modern research into the *non-physical and spiritual* aspects of reality over the last 150 years. Heartfelt thanks, too, to those who have put up with me on this topic over the years, especially my dearest friend, T.

 First published in Great Britain in 2026 ISBN: 978-1-78963-647-5

Although, due to resource limits, replies cannot be guaranteed, the author can be messaged at isevolutiontrue@proton.me. There may, in due course, be further information and resources available at isevolutiontrue.net, although, at time of publication, this is just the seed of an idea.

"Miracles are *not* contrary
to the laws of nature but only contrary to
what we *know* of the laws of nature." St. Augustine.

"The belief that an organ so perfect as the eye could have
been formed by natural selection is enough to stagger anyone."
Charles Darwin

"... the black bear was seen... swimming for hours with widely open mouth, thus catching, like a whale, insects in the water. Even in so extreme a case as this, ... I can see no difficulty in a race of bears being rendered, by natural selection, more and more aquatic in their structure and habits, with larger and larger MOUTHS, till a creature was produced as monstrous as a whale."
Charles Darwin

"One cannot look at this Universe with all [its] living productions and
man without believing that all has been intelligently designed;
yet when I look to each individual organism,
I can see no evidence of this."
Charles Darwin

",,,when it comes to the often
gross extrapolations between observations
and conclusions on macro-evolution, scientists, it seems
to me, permit unhealthy leeway. ... when will we cry out, "the
Emperor has no Clothes!" ... I simply do not understand, chemically,
how macro-evolution could have happened." Professor James M. Tour

Nullius in verba, "take nobody's word for it," motto of the Royal Society, founded in 1660. It denotes the Society's sacred commitment to evidence, observation, and experimentation over a slavish reliance on authority.

"Darwin's theory of evolution will be one of the greatest jokes in the
history books of the future. Posterity will marvel that so very
flimsy and dubious an hypothesis could be accepted
with the incredible credulity it has."
Malcolm Muggeridge

Preface and Your Author's Journey

There are only two reasons to question Darwin's world famous theory of Evolution: (1) because it's *not true,* a good reason, or, (2) because we *don't like it,* a bad reason. So, let me assure you, I don't dislike Darwin's famous theory as such. If I still believed it to be true, as I once did, my only disagreement would be that the amazing evolution process which he described, was an *intelligently* driven unfoldment, not a totally *mindless* and *atheistic* one as he proposed.

The trouble is, not only do I now think that Evolution, as he described it, could not have been a wholly random process, I do not think it took place at all, random or *intelligently* guided. This is what I will do my best to show in this book.

My own journey is as follows. As a child, I believed in the *spiritual* in the simple way that children do, based on input from school assemblies and some basic religious studies. Later, as a teen, I read about Darwin, and of his atheistic ideas about Life's story, and, as a result, I decided that he had found a viable *God or Spirit* not needed explanation for Life on Earth. So, I became a Darwin admiring atheist.

My new *anti-spiritual* way of thinking continued for a few years, until, in my early twenties, I took up meditation, purely for relaxation, as it happens, not for any spiritual reason. However, as I read deeper into the topic and the ideas behind it, I became more open, once again, to the classical idea that there might, after all, be more to reality than my very pro-Darwin, teenage atheist philosophy had allowed me to believe.

At this stage, I did not doubt Darwin's theory overall, but I *did* begin to question how such a wonderful and complex process could have been driven by pure chance? It seemed far more logical, and likely, that there was some kind of *driving intelligence* behind it all, or, *within* it all.

So, my writing here is guided not so much by some dogmatic belief that evolution cannot be part of reality, but that, when it comes to *Living things,* it must have a *driving, guiding, intelligence* behind it. Nature's Amazing DNA Coded Creations, from tiny Microbes, to Fish, to Snails, to Bears, to Whales, to majestic Eagles, Apes, and People, were, I began to realize, *far too complex* to have arisen by pure chance. This was not just unlikely, it was *demonstrably impossible*.

While I can readily agree that the world's religions have been used, and abused, *by us,* to cause each other so much strife and suffering, I can no longer disagree with them, as I did in my teenage-atheist years, that there is *far more* to magical, *multi-dimensional,* reality than meets the skeptical atheist eye. There is far more going on with *Life and Love and Creativity* than is 'dreamed of' in the totally *materialistic* ways of thinking which have come to so dominate our schools and universities, our sciences and and our philosophies in recent times.

It is this modern spirit of materialist-atheism[1] which this shorter version of my earlier book, ***Is Evolution True?,*** is intended to address.

Why does it matter though? Firstly, because how can we believe that Darwin's theory is true, even if we *want* it to be, if it *doesn't stack up?*

If the data supported it, I would not fight with his theory, overall. I would simply argue that it was an *intelligently* driven process, not random, as he and his *atheistic* followers, like R. Dawkins, propose.

Secondly, it matters because Darwin's theory of a totally mindless form of evolution has caused so many of us to doubt that we inhabit a *multi-dimensional* Cosmos of *Intelligence and Meaning,* of *Spirit and Soul.* No, we are urged to believe that, at base, we are just meaningless cosmic accidents, that *arose* by accident and *evolved* by accident. What, though, if this grim view is not true? What if, far from elevating science, it *lowers* science to think in this totally *Soul and Spirit* denying way?

Lastly, I make few claims for originality. Most of what I say here has already been said by others, often far better than I can manage. But, while our current mainstream science's ultra-materialistic, *Soul-and-Spirit-denying* way of looking at the world still holds such a grim sway, I think that any contribution, no matter how small, to show MATERIALISM to be false, and *Spiritualism* to be true, is worthwhile.

WHY EVOLUTION IS NOT TRUE

Contents

Introduction

How can anyone say that ***Evolution is Not True,*** because doesn't everything evolve? Stars and Planets?, our Ideas and our Technologies? Well, yes, and, no. Because our word, 'evolution,' refers to a series of *connected* changes, where something, which *already* exists, *gradually* changes or evolves into something else. Landscapes gradually evolve. Carts are evolved into Cars, Radios into TVs, Abaci into PCs.

Evolution does not, though, explain how something comes to be, in the *first place*. An ancient single-cell microbe cannot evolve into a multi-celled *Snail, Trilobite, Worm, Spider or Fish* without first existing.

So *how did* the first mysterious microbe come to be? Did it *arise or 'evolve'* from NON-LIVING CHEMICALS in some ancient sea? And, if so, was it by *intelligent, spiritual* means or by *mindless* means?

Then, having first arisen, did it *evolve* from there into all the Life Forms and *Species* we know today, into a great, branching, tree of Life, as Darwin proposed? Or, did the species arise *separately,* due to some amazing, inherent, self-existing, *Cosmic Intelligence,* which ***just is?***

These were the kinds of questions Darwin was trying to answer, when he first put forward Evolution, as an alternative to the ancient idea of a *spiritually* driven *Creation.* He argued that Life did *not* first arise, then evolve from there, due to *Intelligent Spiritual* forces, which ***just Were,*** but due to pure chance which ***just Was***. It was not long before Darwin's view became the *scientific* view. There was no longer anything *spiritual* to Life, it was just another natural phenomenon, and it was no more mysterious or difficult to explain than the wind and the rain.

So, for Darwin's now-so-famous theory of Evolution to ***not*** be true, it must be for one or both of two reasons: (1) It *could* NOT happen. (2) It *did* NOT happen. Conversely, for it to be true, Darwin's supporters need to show us: (a) that it COULD happen and (b) that it DID happen.

To show us that it 'DID' happen is the most important. Because it *'could happen'* is theory. But if Darwin's fans (a) cannot show us how it *could* happen, and (b) if there is NO GOOD DATA to show us that it DID happen, then *we have to* conclude that, despite all the claims for Darwin, and his once novel idea, "EVOLUTION IS NOT TRUE."

But, you may be thinking, what's the problem with Darwin? Didn't he have a good theory? Wasn't it better than relying on the ancient Creation story in the Bible for understanding our origins, and the rest of Life on earth? These are fair questions. I will try to answer them

Firstly, if all Darwin had meant by evolution was, *'change over time,'* there would be no issues. Because change over time is just a fact. Many different Life Forms have appeared and disappeared across vast time on our planet, trilobites, dinosaurs, ancient birds, most long extinct.

But, mere change is not what Darwin meant by evolution. No. He meant a very slow and gradual change from the *non-existent* to the existent, from some mix of DEAD CHEMICALS in some ancient pond or sea to *living* microbes, then to *invertebrates,* like jelly fish, snails and spiders, or *vertebrates,* like fish and amphibia of various kinds, then to dinosaurs, birds, mammals, apes and finally, to people, like you and me.

"So what's wrong with that?" Two things.

Firstly, there is NO EVIDENCE for any such process. I know it's a controversial claim, but it's based on what the data, gathered by the world's best fossil experts *must* lead us to conclude, even if these same experts mostly do not accept the Darwin-negating implications of *their own* findings. Why? Because of their commitment to Darwin's theory overall, and to MATERIALISM, the belief that only the PHYSICAL is real.

On the other hand, why would any of us object to Evolution if, based on the evidence, it could be shown that it was true? Some might object to Darwin's theory on *religious* grounds, but we are not interested in religion as such here, but in *what is actually true* or, at least, most *likely* to be true. Secondly, even if Darwin was right about the ***What*** of Evolution, how can we agree with him on the ***How*** of it? Because, for Darwin, the ***How*** of evolution was a totally *'natural' or 'materialistic'* process, by which he meant a totally mindless or *'god-not-needed'* process. However, if Evolution *did, and does,* take place, Darwin's claim that it was a *no-intelligence-needed* process is unlikely to work.

Darwinian Evolution, if it is true at all, *must* have a *driving intelligence* to it. It is far *too complex* to be the totally mindless kind of process he argued for, albeit, this is our current science'***s*** claim.

In any case, how do we know that *God or Spirit* does not, or cannot, evolve things? According to Judaeo-Christian tradition, we are created in God's image, and we, both, create and evolve things all the time, Carts to Cars, Radios to TVs, Abaci to PCs. So, what's the problem with an *intelligently guided* evolution as a religious or philosophical idea?

No problem at all, dear reader, except the data, including, *above all,* the fossil data, doesn't support Darwin's version of macro-evolution—random or *intelligently guided.* Secondly, it is easy to show that evolution, as Darwin saw it, not only DID NOT happen, it COULD NOT happen. Most of this book is devoted to these two points.

Despite Darwin's persuasiveness in ***On the Origin of Species,*** the *fossils never supported his theory.* There is, also, no data that such an amazing process, as he proposed—the astonishing rise, from DEAD CHEMICALS to *Living* cells of a truly staggering complexity, then the aggregation of these original single-cells into even more complex, multi-cellular plants and creatures, like *Worms and Fish, Spiders and Snails, Bears and Whales, Flowers and Trees, Monkeys and People*—could *possibly* happen by chance, with no *driving intelligence* needed.

Yet, after Darwin first published, in 1859, it became more and more fashionable to adopt his *atheistic* worldview. Influential thinkers like Nietzsche and Marx declared God to be dead—*No Soul, no Spirit, no Intelligence, no Meaning, no Point.* Now we would be free, at last, they said, to create our own paradises [or hells], right here, on Earth, without reference to anything higher or wiser than our own misguided egos.

In the west especially, the religious houses began to empty and there was a precipitous decline in the belief that there is *anything more* to our lives than blind luck. No *Spirit,* no *Soul,* no *meaning,* no *Life after death.* This grim new creed. This is not to say that the old religions, *Shamanism, Taoism, Hinduism, Buddhism, Judaism, Christianity, Islam, Sikhism,* among others, had everything right. No, of course not.

There were many dubious things in those ancient traditions, many of which, sadly, continue to this day.

What we are saying though, is that, *logically* and *scientifically,* we cannot disagree with the old religions *basic belief* that there is *more, far more,* to this amazing existence than the purely PHYSICAL eye can see.

So, what, in a nutshell, is Evolution? It is Darwin's totally materialistic theory of creation, Particles to People, Microbes to Men, by pure, mindless chance or PMC. It is the 'modern' hypothesis that there is no *Spiritual Intelligence* behind the arising of Living things, the *Flowers and the Trees, the Amoebas and the Bees, or, even, You and Me.*

The idea that, due to MERE CHANCE, DEAD, ATOMIC PARTICLES, like CARBON or OXYGEN or NITROGEN, *accidentally* combined to become microbes like bacteria and, eventually, people.

Microbes to Men, by pure, AIMLESS, LUCK, is the essence of Darwin's theory. Is it true? No, of course it isn't—as we will soon see.

What qualifies me, though, your most unknown author, to write anything on these complex subjects? Well, firstly, I have done enough research to be very confident that the promotion of Darwin's theory of evolution was, and it remains, much more of a *propaganda exercise* than it is *scientific* truth. Secondly, I have relied much on the writings of those who are experts in this field and who know far more than I.

Thirdly, so glaring are the problems with Darwin's ideas, it is really very easy, even for non-specialists, to point out their flaws. Yet, you may be wondering, if the failings of Darwin's theory are so obvious, how come more people, including other scientists, do not point this out?

Various reasons. A lot of us have grown up with the idea that Darwinian Evolution is just a fact and that to question it is absurd. Many of us also seem to think that if we abandon Darwin's theory, our only option is to treat the Bible, or some other scripture, as if it is an early but accurate science text book which tells us about not just the ***Why*** of

Creation, but the very ***How*** of Creation, how it *literally* happened, and how it *continues* to happen, *right now, and now, and now, 24/7.*

Others of us don't really care. At least Darwin's theory provided us with, we have been taught to believe, a logical theory of Life's origins, and it *was, and it remains,* far more scientific than trying to treat some ancient religious document as if it is an early but accurate science text.

For many people, the adoption of Darwin's theory, by many of the thinking classes, from the middle of the 19th century, released us from many outdated and oppressive social and moral controls imposed on us by the traditional religious authorities, because, evolution, as Darwin conceived of it, was a totally *mindless* or 'god-not-needed' process.

If true, it implied that none of the traditional religions had anything useful to say about the *creation* and the real origins of Life nor about our place in the universe. We were just ***random animals***, no different, in essence, than any of the other animals, with no *spiritual* destiny, other than to further our own, random, ***selfish-gene,*** animal existences, before disappearing, forever, on the death of our **PHYSICAL** bodies.

While, for many, these grim conclusions, led to a crisis in their *spiritual* faith, others, like Marx's followers, embraced them as a new beginning for humanity. Now that, "God was Dead," it was up to us to create Heaven (or Hell) on Earth, because there was nothing else.

This may be why so many of us, who now think of ourselves as *'scientific,'* find it so hard to let Darwin go. Because, despite his theory being falsified, not once, but repeatedly, as we will repeatedly see, so convinced are we that the traditional religions do not benefit humanity, so convinced are we that Darwin's *atheistic* worldview is true, so convinced are we that the *spiritual* DOES NOT EXIST, and *cannot* exist, that no amount of evidence, no matter how powerful, or logical, seems to allow us to reconsider or to penetrate the carapaces of our now so very made-up minds.

This short book, which is an abridgment of our earlier, ***Is Evolution True?,*** *Why Darwin's Rejection of Intelligent Design No Longer Makes Sense,* is a gentle invitation for us to think and to reconsider.

1 What is Evolution?

Darwin's theory of Evolution, which he first published in 1859, at the height of the Victorian steam age, was that Life was not Created, as most people had, until then, believed, but, rather, it had evolved, over vast time, from NON-LIVING chemicals, by PURE LUCKY CHANCE.

There was, he argued, no *supernatural* input, no *spiritual* input into Life's, until then, *seemingly magical* arisings. Life was not, after all, a *miracle,* something truly *mystical, spiritual, or supernatural* in origin.

No, actually, all it needed to start, and to evolve from there, were some CHEMICALS, a warm little pond somewhere, and some flashes of lightning. Darwin wrote to his friend, Thomas Hooker,

> 'But if *(and oh what a big if)* we could conceive in some warm little pond with all sorts of ammonia and phosphoric salts – light, heat, electricity etc present, that a protein compound was chemically formed, ready to undergo still more complex changes'[2] [and, by pure chemical luck, become a *Living Cell, "the most complex system known to man."* [3]]

Darwin's idea was that no *'god' or 'inner'* cosmic *'intelligence'* of any kind is needed to *create* and to *sustain* Life. Rather, if the 'natural' conditions are right, Life will 'naturally' emerge, and it will 'naturally evolve' from there into all the amazing Life Forms we know of today:

First, single cells, bacteria and algae, then, multi-cell plants and animals, trilobites and ammonites, snails and clams, spiders and fish, bears and whales, eagle's wing and peacock's tail, rats and bats, caterpillars and butterflies, flowers and trees, monkeys and people.

Darwin's theory is referred to in various ways, including: *Survival of the Fittest,* natural variations, which are *Naturally Selected* and lead to the *Descent, with Modifications,* of the later, better adapted, variations.

For example: first, there are single-cell bacteria. These, *tiny*, single-cell microbes then vary and they start to become all the Life Forms we know of today. Some bacteria, it is held, eventually, vary VASTLY enough to become multi-celled Worms. Some will vary VASTLY enough to become the first Snails. Some will vary VASTLY enough to become the first Fish, and so on. All these VAST variations due to pure chemical CHANCE—no *intelligent* input.

The first microbes will VASTLY vary, the 'fittest' random microbe variations will survive, and, gradually, over many variations, new multi-cellular *Phyla* [body plans], and new *Species* will emerge. Everything will descend, by random, chance-based variation, from just one or two, *tiny* ancestral microbes. It seemed a logical scheme—to some.

Yet, dear reader, it takes only a little thought to see that this was a highly implausible process, depending on impossibly unrealistic levels of BLIND LUCK. But, implausible or not, Bacteria-to-Fish-to-Frogs-to-Bears-to-Whales was the idea. [Please see your book's covers].

Prior to Darwin, most people had believed that *Life is Created*—it wasn't just some random thing. It seemed logical that *intelligent* Life had *intelligent* origins. This, the classical view, was reflected in humanity's ancient religious beliefs, and religious creation stories, like the story of the Garden of Eden, and Adam and Eve, in the Bible.

How many believed these ancient religious creation stories to be *literally* true we don't know, but it is likely that they believed them to be *spiritually* true. They symbolized to our ancestors that Life had meaningful, *spiritual* origins. It was not just a cosmic accident, which evolved by accident, as Darwin and other moderns came to propose.

This is the point here: to explore, above all, the battle between the materialist worldview, which Darwin championed, and the *spiritualist* worldview of all those who had and continue to have spiritual beliefs: be they *Shamanic, Hindu, Buddhist, Taoist, Jewish, Christian, Muslim, Esoteric, New Age, Idealist, Spiritualist, or other Personal Beliefs.*

Materialists, like Darwin in his day, and Richard Dawkins in ours, believe that only the PHYSICAL UNIVERSE is real and that *Life* and

Mind are just accidental by-products of a BASIC MATTER, atoms of carbon etc, which is, itself, MINDLESS and DEAD, without *Spirit or Soul.* But, due to a long series of *[very!] lucky* chemical coincidences, some of this MINDLESS, DEAD, matter became *Alive!,* and, in time, able to *think,* to *imagine,* to paint, to make music, to write poetry and so on.

This is Darwinian Evolution's Particles-to-People, Microbes-to-Men, by-pure-chance, theory of Existence. It is, also, the current, [2026], mainstream, *scientific* view. To believe in an *intelligently* guided creation, or in an *intelligently* guided evolution, is, according this temporarily very mainstream view, unscientific and taboo.

In case you think I'm making this up, 90%+ of the scientists in America's highest scientific body, the National Academy of Sciences, the NAS, profess to the *atheistic* stance. Darwin's famous theory of Evolution is the *majority* position, and those few scientists who dare to question it tend to be ridiculed or sidelined. Life is just an accident, which evolved by accident, and Mindless Evolution is *'the truth.'* Any who question it are either academic trouble makers or religious cranks.

Fascinatingly, the promotion, as good science, of Darwin's—always unproven—theory was so successful, that so many of us who, before Darwin, had *spiritual* beliefs, gave them up, because he, we felt, had made our beliefs redundant. As Nietzsche had said, "God was dead."

Marx, Lenin, Stalin, Mao, and so many of their followers too, subscribed to this newly minted view. "God is dead! Now we can create our own paradises (or hells!) right here on Earth, without reference to anything higher than our own ideas and our own egos—no matter how *infernal* the result! Those who dare can even be Nietzsche's Superman!"

Not *all* religious people gave up after Darwin. Some made peace with his ideas by telling themselves that he had simply discovered *God or Spirit's* way of creating. Where, after all was the *Spiritual Law* which said that Creation had to be instantaneous? "In any case," they added, "we are made in God's Image, and *we* evolve things all the time, Carts to Cars, Radios to TVs, Abaci to PCs. So maybe Darwin was right about Evolution, just wrong to think that it was a *'god-not-needed'* process."

Other religious people reacted differently. Not only, they said, was Darwin wrong to believe that Evolution was a *mindless* process, but it did not take place *at all.* No. Instead, they began to treat the Bible, not only as a precious *religious* book, but as if it was an early, *but accurate, science* text, which not only described the ***Why*** of Creation—*Spirit*—but the very ***How*** *of Creation,* literally *how God* or *Spirit Created.*

So, if Genesis said that our Earth was made in just Six, very busy, divine days, then that was *literally* true. If, by reading the timelines in the Bible in a very *literal* way, one arrived, at 6,000 years for Earth's age, as in the 1600s, Bishop Ussher had done, then that, also, had to be *literally* true. These folks are known, today, as young-earth creationists.

A third, spiritually minded group did *not* try to treat the Bible as if it was science. Nor, however, did they believe that Darwin had, without realizing it, discovered *God or Spirit's* way of creating Life.

Why not? Because they did not, and still, do not, agree with his theory—regardless as to whether it was a *mindless* process, as he believed, or a divinely guided one as his *spiritually* inclined followers held. No. Those in this last group not only disagree with Darwin about the ***Why*** of Creation—which, for him, was probably blind chance, not *spirit*—but also as to the ***How*** of Creation, which, for Darwin, was his famous theory of Evolution. In sum, there are, currently, four major,

<u>Theories on the Causes and Origins of Life</u>

(a) A *Spiritually* Caused Creation, which was *instant,* in just six days, a mere 6000 years ago. Bible-Based Young Earth Creationism.
(b) A *Spiritually* Caused Creation which Evolved as Darwin described. Some call this Old Earth Creationism, but with Darwinism added.
(c) A *Spiritually* Caused Creation, which is billions of years old, but one which did *not* unfold or 'evolve' in the way Darwin described.
(d) A totally NON-SPIRITUAL Evolution as proposed by Darwin and other *materialist* or *atheistic* thinkers with similar ideas.

Of the four, we will argue *for* cause (c), and *against* (a), (b), and (d). We will not spend any more time on (a) as the Bible is a *spiritual* book, it is not an early *science* text. Like other religious texts, it may tell us something of the ***Why*** of Creation, about the *deeper meanings* of things, but it can tell us little of the ***How***. Earth is clearly far older than 6000 years, and the idea that all Life was created in one go, in just 6 days, and that it has not changed much since then, does not fit the data.

Our Young Earther friends can be glad, though, that although these facts negate their Young Earth theory, they do not give support, in any way, to Darwin's thinking. An *intelligently* driven creation of some kind makes far better sense of the data than Darwin's ideas; but it has to be Old Earth. Our planet is vastly older than just a few thousand years.

What about (b)? A *Spiritually* Caused Creation, not instant, but one which *Evolved,* much as Darwin said, over deep time? Personally, I would readily *agree* with this view, if it were actually *true,* if the evidence *actually* supported Darwin's theory. Why would I agree?

Because it would support our, humanity's, ancient intuition, and logical inference, that there is *far more* to Life than just the physical, while providing us with the beginnings of a *scientific* understanding of the ***How*** of creation. It was *not instant but gradual.* I cannot, though, believe that Evolution, as Darwin described it, took place at all, either r*a*n*d*0*m*ly, as he believed, or ***intelligently*** guided, as his more religiously inclined followers in group (b) try to believe.

However, while I, unlike teenage me, can no longer believe in Darwin's approach, we will spend most of our time on (d), because, not only is it the current *mainstream-science* view, it is the very view which this book is designed to challenge. To explain why, despite all the claims, Darwin's theory of Evolution does not work and, how, ironically, it is just as easy to refute as the Young Earth Creationist view.

It is very easy to show, as we will soon see, not once but repeatedly, that there is no way that Life, whether we Love it, or Hate it, could arise by PURE CHANCE. However, while Life is far too complex to ***arise*** by pure luck, as Darwin thought might happen, there is no evidence that it ***evolved*** in the random way he said either. The fact that so many of our *text books* now tell us that Darwinian Evolution is just a fact—like gravity, or that our earth is round, not flat—does not make it true.

While Darwin's *atheistic* hypothesis of Life appeals to those who don't like the religions or who, for some other reason, cannot bring themselves to believe in any kind of *Meaningful, or Intelligent, Higher Power,* it does not work, both in the sense that it COULD NOT happen, and, even more importantly, there is NO DATA to show that it DID HAPPEN, despite all the claims for Darwin, since 1859.

This, again, is our reason to write: to question Darwin's once radical idea, his passionately held hypothesis, that Life is just an accident which evolved by accident. His materialist (atheistic) idea that all we, and every fellow *Living Being,* are is just a SET OF CHEMICALS, and when the chemicals stop, we drop, and that's all we ever were.

Of course, if this view were honestly true, we'd just have to accept it and to move on—as many of Darwin's supporters try to get us to do.

"Life and Mind, Awareness and Feelings, Love and Laughter, Curiosity and Creativity," they say, "are none of them anything special. They are all just ACCIDENTS OF CHEMISTRY which became bio-chemistry! All the world's spiritual traditions are just wishful thinking.

"Every *religious* or *spiritual* experience anyone ever had, or claimed to have had, is but a hoax or a delusion.[4] No near death,[5] out of body[6] or other so called *spiritual* experience is ever what it *seems* to be…

"No apparently meaningful coincidence was ever truly meaningful. No, it was just a random fluke. Sorry, spiritual people, the harsh truth is that—if we just give it enough time—the MINDLESS ATOMS of carbon, oxygen, and nitrogen, of which we are all so *purposelessly* made, will, clump together, by *pure, unintelligent, chance,* and, eventually, become you and me. This, is the *scientific* view, the *rational* view, the *logical* view. Now go out and enjoy your meaningless lives!"

Ah, dear reader, we do live in such strange, topsy turvy, back to front, and upside down times. Our ancestors would have laughed at us.

But, perhaps, because we have, now, had so much scientific and technical success, in manipulating the PHYSICAL WORLDS to our benefit, we have lost sight of the deeper causes[7] and deeper meanings of things? We have, in fact, decided that there *are* no deeper causes.

2 Could Life Arise By Chance?

Darwin wrote to his friend, Thomas Hooker, "But if *(and oh what a big if)* we could conceive in some warm little pond with all sorts of ammonia and phosphoric salts – light, heat, electricity etc present, that a protein compound was chemically formed, ready to undergo still more complex changes'[8] [and, by pure chance, become the first *Cell(s)*.]

When I was younger, there used to be an urban legend: "No need to believe in any kind of *Higher Power,* because the *laws of probability* can create pretty much anything. Just fill a room with some monkeys and PCs, let them type randomly, and they will, eventually, produce the complete works of Shakespeare, such is the power of probability!"

After I first learnt about Darwin's ideas, which led me to become a very pro-Evolution, teenage-atheist, I too used to trot out this silly urban legend, as a way to make the point that Life could, also, arise by pure chance, which was, of course, the point of the story.

However, in *Genesis and the Big Bang,* physicist Gerald Schroeder considers the odds of monkeys randomly typing out even just one, brief, Shakespeare poem, *Shall I Compare You to a Summer's Day?* He writes,

> 'There are 488 letters in the sonnet . . . The chance of randomly typing the 488 letters to produce this one sonnet is one in 26 to the 488th power, or one in 10 to the 690th power. The number 10 to 690 is a one followed by 690 zero's! . . . [Yet] since the Big Bang, 15 billion years ago, there have been only 10 to the 18th power number of seconds . . . [At] one random try per second, with even a simple sentence, having only 16 letters, it would take 2 million, billion years (the universe has [only] existed for about 15 billion years) to exhaust all possible combinations.'[9]

An article titled *Infinite Monkey Theorem,* makes a similar point:

'Even if *every* proton in the observable universe were a monkey with a typewriter, typing from the Big Bang until the end of the universe ... there is *less* than a one in a trillion chance of success that such a universe made [entirely] of monkeys could type any particular document A MERE 79 CHARACTERS LONG.' [10]

Yet, today, when it comes to the origins of *Living Things,* all far more complex than any poetry, *we are still urged* to believe that Life is a mere chemical fluke, that this is the *'scientific'* view, the *'rational'* view, the *'logical'* view. 90%+ of scientists in America's National Academy of Sciences, NAS, profess to atheism. Can they all be wrong?

Maybe lots of monkeys, typing forever, could never type anything meaningful, but does that mean that *Life* could not start by chance?

Let's see. The distinguished astronomer, Sir Frederick Hoyle, wrote:

"Life *cannot* have had a *random beginning*... The trouble is that there are about two thousand enzymes, and the chance of obtaining them all in a random trial is only one part in $10^{40,000}$ [1 followed by 40,000 zeros] an outrageously small probability that could not be faced even if the whole universe consisted of organic soup.[11]"

In response to this, Eastman and Missler write,

'Hoyle's calculations may seem impressive, but they *don't even begin* to approximate the difficulty of the task. He only calculated the probability of the spontaneous generation of the proteins in the cell. He did not calculate the chance formation of the DNA, RNA, nor the cell wall that [protects the cell].

'A more realistic estimate for spontaneous generation has been made by Harold Morowitz, a Yale University physicist. Morowitz imagined a broth of living bacteria that were super-heated so that all the complex chemicals were broken down into their basic building blocks. After cooling the mixture, he

> concluded that the odds of a single bacterium *re-assembling by chance* is 1 in $10^{100,000,000,000}$. This number is so large that it would require several thousand blank books just to write it out. To put this number into perspective, it is MORE LIKELY that you *and* your entire extended family WOULD WIN the state lottery EVERY WEEK, for a MILLION YEARS, than for a bacterium to form by chance!'[12]

Biologist Michael Denton makes a similar point in a different way.

> "Although the tiniest bacterial cells are *incredibly* small… each is in effect a veritable *micro-miniaturized factory* containing thousands of *exquisitely designed* pieces of intricate molecular machinery, *made up altogether* of one hundred thousand million atoms, *far more complicated* than any machinery built by man and *absolutely without* parallel in the non-living world." [13]

Surely, dear reader, if Darwin and his steam-age friends had known of these things, would they not have revised their views? Surely they would have thought it very strange to continue to maintain, in light of all our knowledge today, that such sophisticated processes might arise by *pure, 'chemical luck,'* with no more organization needed for them to arise than for the kinds of processes involved in basic physics and geology, as Darwin had imagined might be possible. I trust you agree.

The point is, Darwin and his friends had no idea how just how *complex* living things would, finally, turn out to be. But we *do.* We have no excuse to continue to claim that, "But if *(and oh what a big if!)* we could conceive in some warm little pond with all sorts… [of chemicals] – light, heat, electricity etc present,"[14] *Life* could arise by chance.

The trouble is, what started out, in Darwin's letter, as a very hypothetical *'if,'* morphed, over time, into one of the most ***unreal*** dogmas of modern science, *despite* the total lack of empirical support for it. Why? Because, in Darwin's day, it was thought that the kind of Life which might play the role of his hypothetical microbial-ancestor-to-all-later-life, was relatively *simple*. This is why he wondered whether cells might not arise *naturally*—by pure luck.

Darwin's contemporary, Haeckel, called the Cell a *'homogeneous globule of protoplasm.'* Pretty simple, you see. So, why shouldn't cells arise by *chance?* Darwin and Haeckel were, however, very mistaken.

Cells, they did not realize, are astonishingly complex. They could *never* arise by chance. Although tinier than sand grains, every cell is *far more complex,* at its own, *within-the-cell* level, than the much larger organs, like lungs, livers, or kidneys, it helps to build.

For example, we, the very clever human people, can nowadays build heart-and-lung or kidney machines. Let's think this through: we dig raw, rough rocks and oils from the ground, then, we turn them into these very clever machines. This is a *stunningly clever* thing to do.

But, the idea that we could build even *just one, 24/7 dynamic, Living, self-reproducing, self-repairing Cell, from scratch,* one tiny cell, *vastly more complex* than any heart, lung, or kidney machine we make, or, indeed, anything we make, would be laughable. Darwin and his steam age colleagues had no idea that, as Denton outlines in his important and influential 1985 book, *Evolution a Theory in Crisis,*

> 'To grasp the reality of life as it has been revealed by molecular biology, we must magnify a cell 1000 million times until it is ***20 km in diameter*** and represents a *giant airship* large enough to cover a great city like London or New York . . . On the surface of the cell we would see millions of openings, like the portholes of a ***vast spaceship,*** opening and closing to allow a continual stream of materials to flow in and out. If we were to enter one of these openings we would find ourselves in a world of supreme technology and *bewildering complexity*.
>
> 'We would see endless highly organized corridors and conduits branching in every direction away from the perimeter of the cell, some leading to the central memory bank in the nucleus and others to assembly plants and processing units.

'The nucleus itself would be a vast spherical chamber more than a ***kilometer in diameter***, resembling a geodesic dome inside of which we would see, all neatly stacked together in ordered arrays, the miles of *coiled chains* of the DNA molecules. A huge range of products and raw materials would shuttle along all the manifold conduits in a ***highly ordered*** fashion to and from all the various assembly plants in the outer regions of the cell.

'We would wonder at the *level of control* implicit in the movement of so many objects down so many seemingly endless conduits, all in perfect unison. We would see all around us, in every direction we looked, all sorts of ***robot-like machines*** ...

'We would see that nearly every feature of *our own advanced machines* had its analogue in the cell: artificial languages and their decoding systems, ***memory banks*** for information storage and retrieval, elegant control systems regulating the automated assembly of components, error *fail-safe and* ***proof-reading devices*** used for quality control, assembly processes involving the principle of prefabrication and modular construction . . .

However, it would be a factory which would have one capacity not equaled in any of our own most advanced machines, for it would be capable of *replicating its entire structure* within a matter of a few hours ...' [15]

Each! of the—*trillions!*—of cells in our bodies, is, at its own level, *like a vast, micro-miniaturized, super-computer-controlled chemical factory, far more complex* than any of our factories or even whole groups of them. *Cells* are *astonishingly, mind-blowingly complex!*

Unfortunately, once some of our earlier modern thinkers began to realize that the old religious creation stories were not *literal* descriptions of natural history [not, surely, that they were ever intended to be?] they threw out the baby of humanity's ancient intuitions of basic *spiritual* truths with the bathwater of outdated religious viewpoints.

As a teenager, after first meeting with Darwin's ideas, I, too, went through this far too simplistic way of thinking. I began to think that all *spirituality* was make believe, and that, *if we couldn't see it* or measure it *physically*, it couldn't exist. I thought I was being ever so 'logical' and 'scientific.'

I did not yet realize that there are many aspects of *Life* that cannot be reduced to physical measurements alone. That there is, not just *emotionally*, but *logically*, a role for *meta-physics*, for the *supernatural* and for the *spiritual* in our understandings of Life, as, when we consider qualities like *Sentiency and Mind, Love and Joy, Curiosity, and Creativity*. Like the *Life Forces*, which the Chinese and Indian traditions refer to as *Chi or Prana, and, especially*, *A W A R E N E S S, S E N T I E N C Y, C O N S C I O U S N E S S.*

Currently, our science denies that *Consciousness* and *Mind* can be *inherent* principles of reality, as *ontologically* real. Rather, it affects to see C O N S C I O U S N E S S and M I N D as accidental byproducts of a **MINDLESS ATOMIC MATTER**, which, it says, is the only '**STUFF**' that's truly real. This is why, according to this way of thinking, if something, like *Chi, or Prana, or the Life Forces, or Spirit, or the Soul,* cannot be *physically* detected, it cannot exist.

Yet, please note, that for our current, totally *materialistic* story on Life's origins to *seem* true, we have to believe so much *more* non-sense than was ever asked of us by the religions. Because, despite their many failings, at least the religions never insisted that Life is just a long series of *impossibly* unlikely chemical accidents, without purpose or meaning.

This strangely *unscientific* way of thinking which invites us to agree to the impossibly implausible proposition that *pure chance* is a supremely powerful creative force, even as it asks us to dismiss every single *spiritual, mystical, or psychical experience*[16] anyone has ever had, just to try to make its own, very ill-thought-through, schemes seem to be true.

How many of us ever stop to think that for the current, totally materialistic or anti-spiritual account of Life's origins to seem *'true,'* we must agree that every *Living thing,* every *Living Being, our Children, our Pets, the Flowers and the Trees, the Birds and the Bees,* is, *essentially,* just a lucky chemical fluke?

That it *literally* arose, and it *continues to arise, right NOW!, this amazing NOW!, and every amazing NOW!,* not for any intelligent reason or meaning, but due to *pure chance?*

How many of us ever stop to consider that for the current, Darwinian approach to seem *true,* we have to agree that *Life* has no intelligence, no meaning, no point.

Yet, ironically, materialism, as an *ideology,* is self-cancelling. Because, IDEOLOGIES like materialism or *naturalism* or *atheism* are not **PHYSICAL** things, are they? They arise from the abstract realm of the MIND and the SPIRIT.

While our *abstract* IDEAS can be represented by symbols, and turned from *thoughts into poems, paintings, or equations,* represented by physical Ink on **PHYSICAL** paper, or into physical Houses, Cars, TVs, and PCs, in *essence* they are *non-physical, mental, spiritual, ideational.*

What, then, you may be wondering, has been the response of the mainstream scientific community to all the data which falsifies materialism, including Darwin's ideas? If you answered, "To ignore the data?," you are correct. But why ignore the data? Because, right now, the mainstream science community's *belief* in **materialism** comes first.

As the biologist, Richard Lewontin, expressing this view, writes,

> "Our willingness [as scientists] to accept scientific claims that are *against common sense* is the key to an understanding of the real struggle between science and the supernatural.
>
> "We take the side of science in spite of the *patent absurdity* of some of its constructs..." [Life is just a chemical fluke,] "in spite of the tolerance of the scientific community for

> *unsubstantiated just-so stories* [Darwin's scheme], because we have a prior commitment, *a commitment to materialism"* [to blind, aimless, chance as a supreme creative principle] ...
>
> "Moreover, that materialism is absolute, for we cannot allow a Divine Foot [intelligent causation] in the door. ...To appeal to an omnipotent deity [to intelligent causation] is to allow that at any moment the regularities [laws] of nature may be ruptured, that Miracles may happen."[17]

What Lewontin says, with such conviction, is not science but *faith.* Does he *know* that materialism is true? Has science 'proved' it to be true? Not when I last looked. He does, though, very much express the current, mainstream, science view, where it is not at all acceptable to believe in—or to argue for—intelligent design. No, far from it.

Currently, for a theory on life's origins to be 'scientific,' we are urged to believe that pure chance or unintelligent design can build cells and shells, hearts and lungs, blood and brains, kidneys and veins, rats and bats, bears and whales—all for no rational or logical reason at all.

But, we must please ask ourselves, is this way of thinking honestly rational? If we can't PHYSICALLY see Life's deeper causes, we cannot, infer them using classical logic and common sense? Lewontin says no.

What Lewontin, and all his similar thinking colleagues, also miss, though, is that the very existence of *Existence,* the very *Being of Being* is the most amazing miracle of all. Why? Because how can there be *any-thing* or any-*Being* at all?, even just one atom, let alone an Entire Universe, let alone whatever ***It IS*** which Causes that Entire Universe, from Quark to Galaxy, and everything in between?

Yet, while we may never be able to know ***why*** there is a Cause of anything at all, versus nothing at all, we can, at least, ask, is it more likely to be a 'dumb' miracle of a Cause which ***just is,*** or an *intelligent* one? Is the rhythmic beating of our hearts, the ceaseless circling of our blood, the constant repairs of our bodily systems, 24/7, more likely to be created by a totally Mindless Cause which ***just is,*** or an *intelligent* one?

Which, dear reader, do you think is more likely to be true?

3 Cosmic Fine-Tuning

Lewontin's proposal that 'science = materialism' is, ironically, ***not*** good for science. Because it assumes we "know" materialism to be true, when this is the very question at issue. Is it true? Is the visible, PHYSICAL universe the only dimension? Or, are there other, *non-physical, spiritual* dimensions? Is reality mono-dimensional or is it *multi-dimensional?*

Most of our reasons for denying that the Cosmos which births us is *intelligent and meaningful* are not scientific, they are emotional.

However, both MATERIALISM and *spiritualism* are hypotheses about the true nature of reality which, thank goodness, *can* be put to the test scientifically. It is good, because we *should* test our claims. But, because science has, especially since Darwin first published, been so captured by the materialist (atheistic) worldview, its *believers* tend to take no notice of the, by now, huge mass of data which *falsifies* MATERIALISM.

What, though, if we present to them the fascinating modern data of the cosmic fine tuning of which Darwin knew little or nothing?

Physicist Paul Davies tells us that, following the Big Bang, "if the ratio of the nuclear strong force to the electromagnetic force had been different by 1 part in 10^{16} [1/10,000,000,000,000,000], no stars could have formed," and, "the ratio of the electromagnetic force-constant to the gravitational force-constant must be equally delicately balanced. Increase it by only one part in 10^{40} [1 in 100,000,000,000,000,000,000, 000,000,000,000,000,000,000] and only small stars can exist; decrease it by the same amount and there will only be large stars." He adds,

"You *must have* both large and small stars in the universe: the large ones produce elements in their thermonuclear furnaces; and it is only the small ones that burn long enough to sustain a planet with life.' [18]

This is an astonishingly precise condition, but how can those of us who are not astrophysicists, grasp just how *stupefyingly* exact it is? Distinguished Oxford Mathematician, John Lennox's account of this illustration, suggested by the astrophysicist, Hugh Ross, is instructive. Lennox writes, "Cover America with coins in a column reaching to the moon (236,000 miles away), then do the same for a BILLION other continents of the same size. [Then] Paint ONE COIN RED and put it somewhere in ONE of the billion piles. Blindfold a friend and ask her to pick it out. The odds are about 1 in 10^{40} that she will."[19]

Will this convince our materialist friends that there may, after all, be more going on in existence than meets the materialistic eye?

The astronomer, Sir Frederick Hoyle wrote, "Once we see… that the probability of life originating at random is so utterly minuscule as to make it *absurd,* it becomes sensible to think that the favorable properties of physics on which life depends are in every respect deliberate…

"It is therefore *almost inevitable* that [this] must reflect… *higher intelligences* even to the limit of God… such a theory is *so obvious* that one wonders why it is not widely accepted as being *self-evident.* The reasons are *psychological* rather than scientific." [20]

So, why have so many of us adopted the *atheistic* worldview despite the *scientific*—not religious—weakness of our position? Why do we tolerate Darwin's unsubstantiated *'Just So'* stories? Is it because we have a philosophical commitment to materialism, as Lewontin urges us to have? Is it because we cannot—PHYSICALLY—see the *spiritual?*

Or is it because the amazing *technical successes* of our sciences, in recent times, led more and more of us to become skeptical about our old religious stories of the *spiritual* and the *supernatural?* As we made ever more *scientific* sense of things, did we begin to believe that the only kind of knowledge that really counted was PHYSICAL knowledge? And that the only laws that truly existed were the laws of PHYSICAL nature?

Is this why we stopped believing in any *Inner or Higher Spiritual Nature* which is ***just as real*** as outer, physical, nature? And we began to forget, that, actually, it is this *deeper, Spiritual Nature* which is the *ultimate cause* of PHYSICAL nature herself? We began to think that if we could see it, or measure it, it was REAL, if not, it was not! And those who answered, "But we cannot PHYSICALLY see or weigh

Consciousness or Awareness, Mind or Intelligence, Love or Joy, Sadness or Fear, Curiosity or Creativity," began to be ignored.

Was it because our scientists and engineers, crowned by so many successes, in so many *practical* fields, were in no mood to give way, once more, to the *priests* and *theologians,* who, they felt, had little more to offer us than ancient stories, of far off events, which could not now be verified—like people performing miracles or returning from the dead.

What, in any case, had our old religious leaders given us? They had endlessly told us how to behave, while, often, behaving very differently themselves. They promised us hope in a Life to come, yet, in *this* world, they cozied up to the high political powers of their day, while often ignoring the plight of the weak and the less-advantaged.

This is partly why, after Darwin put forward his theory of Evolution, versus a *Spiritual Creation,* so many people gave up on their former religious beliefs, and concluded, as I too once did, that he had made all our prior beliefs in a *Higher Power(s)* redundant. Life was just a chemical accident, which evolved by accident, and that was that.

Others, realizing that their scriptures were *not* early science texts describing not just the ***Why*** of Creation *(God, Spirit)* but also the ***How*** of Creation, kept to their scriptures for the ***Why***, but opted for Darwin for the ***How***. Creation was no longer instant, no, it was gradual and it was evolutionary.

Unfortunately, our *current* science of origins, founded, to a great extent by Darwin, has been so taken over by the *scientific-materialist* worldview that our *scientific-spiritualist* friends have been sidelined.

Yet, science is meant to be an *open minded* quest to find out what is true. Not to be dogmatic, but to ask questions and to test hypotheses: Is the universe just PHYSICAL? Or does it consist, also, in *Spirit dimensions?* Is it mono-dimensional or *multi-dimensional?*

Is it *meaningful* or MEANINGLESS? To assume that we have already answered these questions, as Lewontin urges to do, is a mistake. But if we start by assuming materialism to be right, then we *have* to confine ourselves to a purely MATERIALISTIC *ontology,* even if it is false.

These are the battle lines. Not, is it more 'scientific' to be Christian, Hindu, Buddhist or Muslim, but is it more scientific to be a *scientific-materialist* or a *scientific-spiritualist?* The scientific-materialist says,

"Only the PHYSICAL is real, there is nothing more. If we cannot see it in our telescopes or microscopes, if we cannot weigh it or measure it, it cannot exist." The *scientific spiritualist* says,

"Based on all the data we have—the data of our own *intelligences,* the stunning *complexities* of Living things, the cosmic fine tuning, the testimony of countless, everyday people to a vast range of *spiritual, mystical,* and *supernatural* experiences,[21] out of body,[22] near death,[23] telepathy,[24] pre-cognition, clairvoyance, medium-ship,[25] evidence for reincarnation[26] and much more—we think it makes much better sense, overall, to hypothesize that, despite Life's deeper sources *physical* invisibility, we emerge from an Intelligent Cosmic Mystery which ***just Is,*** not a mindless Mystery which—no less mysteriously—***just Is.***"

These are our *scientific,* not religious, choices: we either emerge, without meaning or soul, from a *dead, spiritless* matrix generated by a Mindless, Unintelligent and Accidental Universe Generating Mechanism, MUA-UGM, which seemingly 'impossibly,' ***just Is,*** or we ***meaningfully*** emerge from an Amazing, Ultimate, Miracle of a Universe Generating Mechanism or an AUM-UGM which, also, very mysteriously, and, seemingly 'impossibly,' ***just Is***.

Why *'impossibly?'* Because how is it possible that there is a CAUSE of anything at all, versus nothing at all? Yet, the seemingly impossible has happened, atoms exist, planets exist, stars exist, we exist.

All these things have causes. What caused the Big Bang? What gave rise to Nature's amazing Laws? Or Protons, or Plants, or People, or Planets, or Stars, or Galaxies? Logically, there must be some *Final, Uncaused Cause,* which ***just Is,*** behind all these things. Yet, *how* can such an *Ultimate, Uncaused Cause,* which ***just Is,*** exist? What could cause *it* to—so *mindlessly,* or, so *blazingly intelligently* ***just be****?*

We will never know. Some say, "It's the Big Bang, and everything flowed from that." Ok, but what caused the Big Bang? And what caused that? And that, before that, on and on, the same question repeating for ever. So, I trust you'll agree, we'll never be able to say ***why*** there is a cause of any-thing at all, rather than no cause at all.

What we *can* try to answer though, is which of the two hypotheses as to the *Ultimate Nature of the Final, Uncaused, Cause of All,* which ***just Is,*** is the more logical and the more intellectually fulfilling, overall?

Is materialism's MUA-UGM which *just mindlessly* ***is,*** the better explanation for Nature's amazing Laws, for Life's DNA Coding, "like a computer program, but far more advanced than any software we have ever created,"[27] for Cells and Shells, for Blood and Lungs, for the Lion's Mane, and the Human Brain, for Bats and Snails, Bears and Whales?

Or, is the classical *spiritualist* hypothesis of the AUM-UGM which ***just Intelligently Is,*** the better explanation? Most scientists, today, like Darwin in his day, go for the MUA-UGM. When they look in the mirror, they do not see an *amazing, Living, Soul,* looking back at them.

No, what they *claim* to see is a Soulless accident of *chemistry.*

Presumably, this is also how they see their beloved *Children?,* and their *Pets?,* their *Friends?,* their *Family?* This is the modern materialist-atheist[28] view. They say it's the logical view, the rational view, but is this true? Does this commitment to materialism honestly make sense?

The materialist philosopher, Daniel Dennett, thought it was true. He wrote, "We're all zombies. Nobody is conscious."[29] Francis Crick, discoverer of DNA wrote, "You, your joys and your sorrows, your memories and free will, are in fact no more than the behavior of a vast assembly of nerve cells and their associated molecules."[30] Richard Dawkins writes, "In a universe of electrons and selfish genes…you won't find any rhyme or reason in it, nor any justice… no design, no purpose, no evil, no good, nothing but pitiless indifference,"[31] how grim.

A reviewer of Chris Carter's book, ***Science and the Near Death Experience: How Consciousness Survives Death,*** writes, "What is the job of science? Is it to explain things or to *explain them away*, following a pre-set belief system like materialism, which we do not question?"[32] But, he says, if we do not question we are no longer practicing science

4 Cosmic Soup

One of the modern, origin of Life myths many of us grew up with is the idea of a pre-life CHEMICAL soup sea, or pond, which, when the conditions were right, led, it is said, to Life arising—by chance—as in Darwin's "warm little pond" idea. Except, you may be surprised to learn, as I was surprised to learn, there is *no evidence* for any kind of pre-life, chemical-soup pond or sea. Biologist, Michael Denton, writes,

> "Rocks of great antiquity have been examined ... and in *none of them* has any trace of abiotically produced organic compounds been found . . . Considering the way the pre-biotic [pre-life] soup is referred to in so many discussions of the origin of life as an already established reality, it comes as *something of a shock* to realize that there is absolutely *no* positive evidence for its existence." [33] [No evidence.]

Eastman and Missler add, "The destructive effect of oxygen and ultraviolet radiation from the sun ... makes it *unlikely* that significant quantities of viable nucleotides and amino acids could ever accumulate in the primitive ocean." However, "even if they did accumulate... the next step is to explain *how they combined* [by chance] to form the self-duplicating DNA molecule and the *thousands of proteins* found in the simplest living cells. For the materialist [atheistic] scenario to be taken seriously, it must provide a plausible explanation for the origin of these enormous molecules *without* the introduction *[by an intelligence]* of biochemical knowhow..." They continue,

> "One of the most difficult problems for the materialistic scenario on the origin of life is something called molecular chirality. The building blocks of DNA... are molecules which can exist in both right and left-handed mirror-image forms.

This [left and right] “handedness” is called “chirality.” . . . In all living systems the building blocks of the DNA and RNA exist exclusively in the right-handed form, while the amino acids in virtually all proteins in living systems, with very rare exceptions, occur only in the left-handed form.

“The dilemma for materialists is that all “spark and soup-like” experiments produce a mixture of 50% left (levo) and 50% right- handed (dextro) products. ... [Yet,] Unfortunately, such mixtures are ***completely useless*** for the spontaneous generation of life.

“For 80 years chemists have been trying to synthesize optically pure mixtures of amino acids in the lab using stochastic [random] chemical processes. However, this has never been accomplished. According to physical chemists, it is IMPOSSIBLE ... Miller and Urey acknowledged that the chemical makeup of their [spark-and-soup] experiment consisted of *equal* portions of left-handed and right-handed amino acids.

“Complex molecules such as DNA and proteins are built by adding one building block at a time onto an ever-growing chain. In a [hypothetical] “primordial soup” made up of equal proportions of right and left-handed building blocks, there is an equal probability at each step of adding either a right or left-handed building block.[34]

“Consequently, it is a MATHEMATICAL ABSURDITY to propose that *only* right-handed nucleotides would be added time after time without a single left-handed one being added to a growing DNA molecule. Sooner or later an incorrect, left-handed nucleotide will be added. The same goes for proteins…Consequently, *if even one* nucleotide or amino acid with the incorrect “handedness” is inserted into a DNA or protein molecule, the three-dimensional structure will be annihilated and it will cease to function normally.”[35]

So, if the idea that Life could arise by chance is a mathematical absurdity, how come anyone still believes in it? Emotional reasons? The dogma that Life arises by chance? Yet, pure chance cannot give rise to *two key* features which distinguish all living things, and all the mechanisms we make. Any Car, TV, or PC must be assembled in a **specific** (non-rA*n*do*m*) order or it won't work. Mere chance cannot give rise to this kind of non-random complexity. The laws of probability forbid it. This is why, if we randomly tumbled Car, PC or TV parts around, for as long as you like we'd *never* get a working machine.

The mix would be *complex* but *non-functional.* Any natural ordering would, at best, cause the heaviest parts to sink to the bottom of the mix.

As Stephen Meyer has pointed out, REGULAR NATURAL laws, on their own, do not have the necessary creative power, nor any reason, to generate the *totally specific sequences* of parts and software which characterize all computer CODES, all machines, and all Living things.

Chemical coincidences in nature can only give rise either to r*A*nd0m c0m*P*lexity, as in a tornado, or to *simple, specific patterns,* as in crystal formation. Patterns which can be described by simple algorithms like ABC + ABC + ABC. As origins of life researcher Lesley Orgel writes,

> 'Living things are distinguished by their specified complexity. Crystals such as granite [although specific] fail to qualify as living because they *lack complexity;* mixtures of rAnD0m polymers [although *complex*] fail to qualify because they *lack* specificity.' [36]

A regular physical law which can generate simple, regular, specific sequences, like ABC + ABC + ABC, cannot generate complex sequences like TURN + THE + SWITCH + ON + HERE. A sequence like this is *physically indeterminate*. Due to its specific—yet *irregular* content—it cannot be generated by regular physical laws alone. Such a sentence can only be created by *intelligence*—be its source subtle or be it obvious.

Similarly, a DNA CODED sentence like ATAGGAGC + CAACGGTT + CCTTGTAC cannot be determined by natural laws alone, because it is *complex, it is specific,* and it is i*Rreg*U*lar. No **CODE** or software can ever be produced (i) by chance alone (ii) regular laws alone or (iii) some mix of the two. The laws of probability forbid it. Really, this is very obvious—unless, for some reason, we don't want it to be.

Interestingly, some of the more agile intellectuals of our time *are,* like *Sleeping Beauty,* beginning to awaken, and to ask whether materialism's paradigm of blind, unintelligent, design, actually works?

For example, Antony Flew, who, after decades of a very public atheism, decided to follow the evidence as described in, ***There is a God: How the World's Most Notorious Atheist Changed His Mind.*** [37]

What caused him to have a rethink? For one, Life's **DNA CODES,** which, he realized, could never arise by chance. **CODES** are ABSTRACTIONS—they are not **'material,'** are they? They contain IDEAS, INFORMATION, INSTRUCTIONS. The DNA **ink**, made of amino acids, A.C.T. & G., is not the *abstract CODE.*

So it's as though a computer language, like JAVA, was called Silicon Code or SC after the **physical materials** it is conveyed by? Yes. This is why some have suggested, "why not let DNA stand for "Designed Natural Algorithms."[38] Because CODES always consist in meaningful information which can be conveyed in many forms. Even Richard Dawkins, world famous for his atheism, acknowledges this. He writes,

> "What has happened is that genetics has become a branch of information technology. It is pure information. **IT'S DIGITAL INFORMATION.** It's precisely the kind of information that can be translated digit for digit, byte for byte, into any other kind of information and then translated back again." [39]

And this is "precisely the kind of information" which only ever has *intelligent* causes, and ***can*** only ever have *intelligent* causes. It is not r*A*n*D*Om (meaningless) information, sometimes referred to as Shannon Information. No amount of randomly tumbling the **DNA** letters (A, C, T, & G) around, will ever create Life's software: **CAACGGTT +**

CCTTGTAC + CCTTGTAC. Mere chance cannot generate even a simple sentence like, PLEASE + FEED + THE + CAT. The random mixing of letters can only generate Shannon type gi*B*ber*i*Sh, nothing functional.

Further, as Michael Polanyi, the distinguished chemist pointed out, *if* the sequences of DNA letters in life's codes were *physically* determined, by nothing more than the REGULAR LAWS of *physical* chemistry,

> "then such a DNA molecule would have no [meaningful] information content ... [Because, to be useful, the sequence of DNA letters] must be as physically ***indeterminate*** [able to be arranged in any sequence, regular or *irregular*] as the sequence of words is on a printed page." [40] [Regular = ABC + ABC + ABC Irregular = A G G A A T G C + G G C C T A T A.]

Ironically, far from being ***un-intelligently designed,*** so sophisticated *are* LIFE'S CODES, there is, now, research into the use of DNA for data storage and computing. *Futurism.com* announced: "Self-Replicating 'DNA COMPUTERS' are set to change everything."... researchers at the University of Manchester are "working on turning strands of DNA into the next basis for computing." Because DNA'S double helix can take two paths simultaneously, it can solve problems quicker. [41]

In the 1950s Miller and Urey tried to show that Life could have started, by chance, in their 'spark and soup' experiments. But all their [intelligently designed] experiments generated were a few amino acids.

Now it's true that amino acids are among the basic building blocks of life. But a few, random, amino acids don't make even the beginnings of a Life Form. Chemicals unguided by *higher-level* forces, will never lead to anything clever. Biologist Michael Denton writes

> "Instead of revealing a multitude of transitional forms through which the evolution of the cell might have occurred, molecular biology has served only to emphasize the enormity of the G A P.
>
> "We now know not only of the existence of a break between the living and nonliving world, but also that it

represents the most dramatic and fundamental of all the discontinuities of nature.

'Between a Living Cell and the most highly ordered non-biological system, such as a crystal or snowflake, there is a C H A S M as vast and absolute as it is possible to conceive."[42]

Denton adds,

"Molecular biology has also shown that the ***basic design*** of the cell system is essentially the ***same in all living systems***, from bacteria to mammals. In all organisms the roles of DNA, mRNA and protein are identical. . . . In terms of their basic biochemical design, therefore ***no living system*** can be thought of as being ***primitive or ancestral with respect to any other system, nor is there the slightest empirical hint of an evolutionary sequence*** among all the incredibly diverse cells on earth. For those who hoped that molecular biology might bridge the gap between chemistry and biochemistry, the revelation was profoundly disappointing." [43]

Carl Sagan, a well known 20th century intellectual, said, skeptically, of some people's *spiritual and religious* beliefs that:

"Extraordinary claims need to be backed up by *extraordinary evidence."* He made a reasonable point. But there is no support *at all* for our current science culture's *quite extraordinary* beliefs in *pure chance* to create anything remotely functional.

Here is another example of the implausibility of the, 'Life is just a chemical fluke,' scenario. Jacques Monod who, ironically, was, himself, a materialist, and originator of the now well known statement that just "chance" (chemical luck) + "necessity" (regular natural laws) could, he said, explain everything in Nature, had to concede that:

'we have no idea what the structure of a [hypothetical] primitive cell might have been. ... the simplest cells available to us for study have nothing "primitive" about them. ... The development of the metabolic system, which ... must have "learned" [how it "learned" he doesn't say] to mobilize chemical potential and to synthesize the cellular components, [by pure lucky chance] poses Herculean problems. So also does the emergence of the [very clever] *selectively* permeable membrane without which there can be no viable cell.

'But the major problem is the origin of the genetic CODE and its translation mechanism. ... The code is meaningless unless translated. The modern cell's translating machinery consists of at least fifty macromolecular components WHICH ARE THEMSELVES CODED IN DNA: THE CODE CANNOT BE TRANSLATED OTHERWISE THAN BY PRODUCTS OF TRANSLATION [emphasis in original]. It is the modern expression of omne vivum ex ovo [Life only comes from Life, so which came first, the chicken or egg?].'[44]

'In *The Naked Emperor: Darwinism Exposed,* Dr. A. Latham writes,

'Far too often I hear experts confidently postulate about the first cell membranes. They know that *the first life* had to have a *protective coat* [cell wall] to hold the DNA and all the incredibly sophisticated molecular machinery inside the bacteria. ... [yet] the actual details of the [cell] membrane must be coded for in the DNA of the cell.'[45] [Michael Behe calls this *irreducible complexity.*]

As Latham the *spiritualist,* and Monod the *materialist,* point out, *how* could the cell's **DNA CODED** software arise, absent the protective cell wall, which the software *inside* the cell **CODES** for? The Cell's wall and the **DNA** within, coding for that same wall, are *irreducibly* complex. They *must* arise together or not at all. Is this true? Yes. Does it point to blind chance? *Seriously?!* Does it point to an *inherent intelligence* behind living things? That makes the best sense of the data.

V. J. Torley says that, at his website, Professor James Tour, one of the ten most cited chemists in the world, who has authored and co-authored more than 480 scientific publications and 36 patents, writes,

> 'Although most scientists leave few stones unturned in their quest to discern mechanisms before wholeheartedly accepting them, when it comes to the often gross extrapolations between observations and conclusions on macro-evolution, scientists, it seems to me, permit unhealthy leeway. On hearing such extrapolations in the academy, when will we cry out, *"The [Darwinian] Emperor has no clothes!"*
>
> '. . . I simply ***do not understand,*** *chemically,* how macro-evolution could have happened. ... Furthermore, when I, a non-conformist, ask proponents for clarification, they get flustered in public and confessional in private wherein they sheepishly confess that they really ***don't understand*** either. Well, that is all I am saying: I do not understand. But I am saying it publicly as opposed to privately.
>
> '*Does* ***anyone understand*** the chemical details behind macroevolution? If so, I would like to ... invite them to meet with me. Lunch will be my treat.
>
> 'Until then, I will maintain that no chemist understands, hence we are collectively bewildered. And I have not even addressed origin of first life issues. For me, that is even more scientifically mysterious than evolution. Darwin never [formally] addressed origin of life, and I can see why he did not; he was far too smart for that. Present day scientists that expose their thoughts on this become ever so timid when they talk with me privately. I simply ***cannot understand*** the source of their confidence when addressing their positions publicly.'[46]

Torley recalls how, in a talk in 2012, Professor Tour said that *no scientist* that he has spoken to understands evolution, including Nobel Prize winners. Here's what he said.

'. . . I don't understand evolution, and I will confess that to you. Is that OK, for me to say, *"I **don't understand** this"?* Is that all right? ... Let me tell you what goes on in the back rooms of science – with National Academy members, with Nobel Prize winners. I have sat with them, and when I get them alone, not in public – because it's a scary thing, if you say what I just said – I say, *"**Do you understand*** all of this, where all of this came from, and how this happens?" Every time that I have sat with people who are synthetic chemists, who understand this, they go *"Uh-uh. Nope."* ...

'I was once brought in by the Dean of the Department, many years ago, and he was a chemist. He was kind of concerned about some things. I said, "Let me ask you something. You're a chemist. ***Do you understand*** this? How do you get DNA without a cell membrane? [protecting the DNA *inside* the membrane.] And how do you get a cell membrane without a DNA? [Which CODEs for the very membrane *outside* which is needed to protect the DNA *inside*.]

'And how does all this come together from this piece of jelly?" We have no idea, we have no idea. I said, "Isn't it interesting that you, the Dean of science, and I, the chemistry professor, can talk about this quietly in your office, but we can't go out there and talk about this?' [47]

As Sir Frederick Hoyle pointed out, there is a lot more at stake here than *useful, practical, everyday,* science. Emotions and convictions are at stake. careers and reputations are at stake.

George Wald, a distinguished biochemist, wrote: "the *spontaneous generation* of a living organism is IMPOSSIBLE. Yet we are here as a result, I believe, of spontaneous generation,"[48] adding, "we choose to *believe the impossible" not* on scientific grounds, but on "philosophical grounds."[49] How is such an *unscientific* stance possible?

Because, dear reader, in the science of origins, we don't have to be *right.* This area of science is far more about *philosophy* than it is about *practical* science with *practical* implications, let alone about proven facts. It *doesn't have* to be *'right.'* No one is going to die if we are wrong. Yet, as Tour alludes, I'm quite sure many scientists *understand perfectly well* that Life could not *start* by chance nor *evolve* by such means. But, in the current climate, they keep quiet about their doubts.

Why? Firstly, because they are *taught* to never refer to the *super* natural in their explanations, based on the idea that the *super* natural is, either, (a) not real, or, (b) if it *is* real, it is not needed to explain anything in 'natural' reality. This is the approach in modern science known as method.o.logical materialism (MM) or natural-ism (N). Secondly, they do not want to put their careers and their reputations at risk, even if, by doing so, they are simply pointing out what *should* be the obvious!

Thirdly, if there is no *Creative Spirit(s) or Intelligent Cosmic Cause(s)* of any kind, guess *who* get to be the new 'gods' of our understandings? Not the *priests* of old, but our dear *scientists*. So, many of them *bluff* and maintain, in public at least, that these problems don't exist. This is why it takes a strong soul, like a Tour, to risk the ridicule of those so committed to the current, *Emperor's New Clothes,* views.

Why, though, this relentless urge to argue for *totally mindless* causes for Life's origins? For Darwin and some of his colleagues, it was partly because of their sadness at nature's harshnesses. How could the gentle 'good' god, of their Victorian Christian beliefs, allow so much suffering, death, and disease? Yet, can we not, just as rationally, choose to notice the *intelligence and the beauty* in Nature, the magic of a child's smile, and the kindness and love so often around us.

Our materialist and atheistic friends do often point to the endless sufferings in this world, and, due to these, they argue *against Intelligence* and *Soul* in nature.

Others, more *spiritually* inclined, point to the incredibly clever ways in which living things are put together, to the miracles of *Life,* to *Love,* to the human *Spirit,* to human *Creativity,* and to the *goodnesses of Life.*

5 Why Evolution is Not True

The real science is not on the side of those who believe Life could *start* by chance. Even if, currently, the mainstream science community does not agree, due to—wait for it—its prior commitment to MATERIALISM.

A commitment based, not so much on the data, but on a dislike of the alternatives. Why, after all, would the science community, now considered the most knowledgeable in society, wish to give way, once more, to our *meta*-physicians, and to our *spiritual*-scientists, who most *physical*-scientists now consider to be scientists of the *non-existent!*

Science, though, is about finding out what is true, it is not about clinging to a particular belief system, like MATERIALISM, just because we don't like the alternatives. However, even if it is now (very) clear that *Life could never start by chance,* we must still ask, could it ***evolve*** by chance? This is where Darwin and his theory really start to come in.

However Life started, Darwin believed that he could explain it from there. Having decided that 'god,' if 'god' existed at all, would never deliberately create so many horrible things in Nature, he wondered whether a more MECHANISTIC explanation might work, a *non-spiritual* explanation, a purely physicalistic or NATURALISTIC explanation.

Darwin had observed that skillful breeders could breed different varieties of pigeons, dogs, and cattle by *intelligently* selecting random natural variations when they arose. What if Nature had—*aimlessly*— done the same thing? What if, when a 'good' variation arose, it had survived in the wild, and, what if, this led, in time, to new species?

People began calling this 'survival of the fittest.' The only problem is that no breeders had ever managed, nor have they since, to breed a completely new species in this way—using *intelligently* guided artificial selection. Even the tiny Chihuahua is, in essence, still a Wolf, it is not a new species. Darwin was not deterred by this difficulty. He asked, what if there were some tiny, single-cell microbes, like Bacteria. What if they varied? What if those variations became the first multi-cellular animals?

But, dear reader, this is, as I'm sure you'll see, a highly unrealistic idea because the levels of change Darwin's theory needed were HUGE!

In any case, microbial life, it is currently believed, dates back about 3.8 billion years on our planet, yet, in conflict with Darwin's scheme, for the next 3.2 billion years there is ***no sign*** of Darwin's imagined process—of descent with HUGE modifications—of new creatures slowly emerging and branching out from those primeval microbes.

Let's imagine it: tiny bacterium turning into a VASTLY LARGER SNAIL. A Snail is 10,000 + times longer than a microbe! Yet, according to Darwin's theory, a microbe could evolve, by pure chance, we are to imagine, into something as vastly larger as a SNAIL, SPIDER, or FISH, the largest of which, a Whale Shark, can be 20 million times longer.

If a bacterium were the size of a pea, the *smallest* Snail would be *50 meters long!,* a Whale Shark would be c. *60 miles long!,* c. 20 million times longer! We need to *properly think through* the likelihood of Darwin's random-chance evolutionary process achieving such a thing.

The likelihood is zero, and, even more important, what *actually happened* totally conflicts with Darwin's predictions. Evolution, as he described it, did not happen. No. Rather, around 600–500 million years ago, the mysterious biological Big Bang known as the Cambrian Explosion took place, and dozens of new multi-cellular animal body plans ***abruptly appeared***, in a relatively short period of geological time, including examples of all the basic animal body plans we know today.

This was NOT, in any reasonable way, Evolution, as Darwin meant it, and there are ***no plausible fossil ancestors*** for these new Life forms in the *pre* Cambrian rocks. There are no signs at all that the many weird and wonderful creatures of the mysterious Cambrian Big Bang, like Waptia, Marrella, and Canadaspis, descended, in any way, shape, or form, from the single-cell microbes of the previous 3.2 billion years.

The Cambrian Explosion is not evolutionary—it's revolutionary. The Cambrian fossils show the *sudden appearances* of many new *Phyla*

[Life Forms]—*without* ancestor fossils as called for by Darwin's theory of tiny bacterium to SNAIL, SPIDER, WORM, FISH, or WHALE.

In *The Naked Emperor: Darwinism Exposed,* Dr. Anthony Latham discusses how even the world famous fossil authority S. J. Gould felt compelled to point out that Darwin's predictions did not match the data.[50] Gould wrote:

> 'Darwin invoked his standard argument to resolve this uncomfortable problem: the *fossil record is so imperfect* that we do not have the evidence for most of life's history. But even Darwin acknowledged that his *favorite ploy* was wearing a bit thin in this case. His argument could easily account for a missing stage in a single lineage, but could the agencies of imperfection really *obliterate absolutely all evidence* for positively every creature during most of life's history? Darwin admitted: 'the case at present must remain INEXPLICABLE; and MAY TRULY BE URGED AS A VALID ARGUMENT AGAINST MY VIEWS here entertained.' *Origin of Species,* 1859. [My emphasis.]

Latham continues, "As already described, we do know of fossils before the explosion but these *completely overturn* the Darwinian model. For 3.2 billion years we have only microbial life... Then, in a period of about 50 million years: the Ediacaran Fauna which *appear unrelated* to the Cambrian fauna followed by the host of body plans of the Cambrian explosion." Latham quotes from Gould's *Wonderful Life:*

> "Thus, instead of Darwin's gradual rise to mounting complexity… Nearly 2.5 billion years of prokaryotic [non-nucleated, bacterial] cells and nothing else, 2/3rds of life's history in stasis at the lowest level of recorded complexity.
>
> "Another 700 million years of the larger and much more intricate eukaryotic [nucleated] cells [like algae] but no aggregation to multicellular animal life. Then in the 100 million year wink of a geological eye, three outstandingly

> different faunas – from Ediacaran, to Tommotion, to Burgess. Since then... not a single new phylum or basic anatomical design [has arisen.]" [51] Gould continues, "you may *want* to read this sequence as a tale of predictable [Darwinian, evolutionary] progress," *non-nucleated* cells, then *nucleated* cells, then multicellular life, Worms, Snails, Spiders, Fish. But, he warns, "scrutinize the particulars," and Darwin's "comfortable story collapses."[52] Why, for example, did life remain at stage one, for 2/3 of its history, if increasing complexity offered such great biological benefits?

There is ***no evidence*** for Darwin's *"slow and continuous rise of complexity,"* microbes to worms, worms to fish, *water-breathing* fish to *air-breathing* land-walkers, dinosaurs to mammals, mammals to apes, apes to humans. As Gould says, and Michael Denton also notes, there is ***no evidence*** for any kind of evolutionary arc[53] from single-cell microbes to the VASTLY LARGER and more complex multi-celled creatures, like snails, worms, or fish that Darwin predicted. Then, in yet more conflict with Darwin's ideas *not a single new* basic body plan has ever arisen.

He predicted that new body plans would emerge out of the earlier ones. Yet there are *fewer* animal body plans today than before, not more.

There are around thirty basic body plans, including: *Cnidaria,* sea anemones, corals, jellyfish. *Mollusca,* snails, slugs, mussels, octopuses. *Annelida,* marine worms, earth worms. *Arthropoda,* around 3.7 million species, including the classes *Arachnida,* spiders, mites, scorpions. *Insecta,* 1.5 to 3 million species, ants, bees, beetles, flies, butterflies.

Crustacea, shrimps, lobsters, crabs. *Platyhelminthes,* flatworms, parasitic flukes, tapeworms. *Nematoda,* round worms, over 80,000 species. *Echinodermata,* including starfish, sea urchins. Lastly, *Chordata,* the basic anatomical design to which we, vertebrates belong. 60,000 species including fish, amphibians, reptiles, birds, and mammals.

The abrupt arrival of so many new animal body plans in the Cambrian *falsifies* Darwin's entire [macro] theory. Darwin wrote:

"The difficulty of understanding the *absence* of vast piles of fossiliferous strata, which on my theory were no doubt somewhere accumulated before the Silurian (Cambrian) epoch is very great. I allude to the manner in which numbers of species of the same group *suddenly appear* in the lowest known fossiliferous rocks." [54]

Some of Darwin's supporters have argued that the *sudden arrival* of so many new species in the Cambrian only *seems* discontinuous, because any precursors to the Cambrian creatures may have been *soft-bodied* and thus lacking in the kinds of bones or shells which could be preserved in fossil form, but, as Stephen Meyer points out in ***Darwin's Doubt: The Explosive Origin of Animal Life and the Case for Intelligent Design,***[55] *plenty* of soft-bodied organisms are preserved in the rocks from *before, during,* and *after* the Cambrian period.

Meyer also explores why random genetic mutations, RGMs, which have, for a long time now, been held to be the very engine of evolution, Darwin did not know about genes, simply do not have the vast creative powers needed to produce these new forms. Casey Luskin writes,

"[Dr Meyer] cites multiple peer-reviewed studies showing that MULTIPLE COORDINATED MUTATIONS would be necessary to produce functional proteins, but these COULD NOT [POSSIBLY] ARISE within [the limited] waiting times allowed by the fossil record. ... [This is] devastating to [all] Darwinian explanations of the origin of complex features." [56]

No, the species arose suddenly, *not* gradually. Darwin argued that the first animal body plan, a *tiny* microbe, could lead to the next body plan, a VASTLY LARGER *10,000 times* LARGER *invertebrate* WORM which could lead to the next body plan, say a *vertebrate* FISH and so on. But, apart from the biochemically unrealistic nature—impossible is a better word—of this idea, the fossils reveal the very OPPOSITE!

The basic multi-cellular body plans *arrived in one go*. Darwin, though, dismissed the Cambrian data by claiming that there was just too little fossil evidence to support his idea. Luther Sunderland explores this problem for Darwin's theory in *Darwin's Enigma.* He writes, "Geology and paleontology [FOSSILS] held great expectations for Darwin, although in 1859 he admitted that [ironically] they presented the STRONGEST SINGLE EVIDENCE AGAINST HIS THEORY. ...

[The] fossils were a perplexing puzzlement to Darwin [and a huge embarrassment for his theory] because they *did not reveal any evidence* of a gradual and continuous evolution of life from a common ancestor, the *proof he needed* to support his theory." [Hence the CAPS above.]

But Sunderland writes, "Although the fossils were an enigma to Darwin [although they *falsified* his theory], *he ignored the problem* and found comfort in the *faith* that future explorations would reverse the situation and ultimately prove his theory correct." Darwin tried to get around the problem by making this claim in *The Origin of Species.*

> "The geological record is *extremely imperfect* and this fact will to a large extent explain why we do not find intermediate varieties, connecting together all the extinct and existing forms of life by the *finest graduated steps"* [Finest graduated steps? This was delusional. No we do not find *any* such fossils at all.]
>
> Darwin added: "He who rejects these views on the nature of the geological record, WILL RIGHTLY REJECT MY WHOLE THEORY." [57] [Here, at least, Darwin was correct.]

Sunderland continues, "Now, after over 120 years of the most extensive and painstaking geological exploration of every continent and ocean bottom, the picture is *infinitely more vivid* and complete than it was in 1859. Formations have been discovered containing hundreds of billions of fossils and our museums now are filled with over 100 million fossils of 250,000 different species. The availability of this profusion of hard scientific data should permit objective investigators to determine if Darwin was on the right track."

So, he asks, "What is the picture that the fossils have given us? Do they reveal a *continuous progression* connecting all organisms to a common ancestor? With every geological formation explored and *every fossil classified* it has become apparent that these, the only direct scientific evidences relating to the history of life, *still do not provide* any [of the fossil] evidence for which Darwin so fervently longed.

"The GAPS between major groups of organisms have been growing even wider and more undeniable. They can no longer be ignored or rationalized away with appeals to the imperfection of the fossil record [as Darwin and his followers have been so prone to do. No, he adds]...

"By the 1970s prominent scientists in the world's greatest fossil museums were coming to grips with the so-called "gaps" in the fossil record and they were cautiously beginning to present *new theories* of evolution that might explain the *severe conflicts* between neo-Darwinian theory and the hard facts of paleontology [the fossils].

"Back in 1940, Dr. Richard B. Goldschmidt had faced the horns of this dilemma-of-the-gaps with his *hopeful monster theory,* the idea that every once in a while an offspring was produced that was a [GAP jumping] monster grossly different from its parents.[58]

"Goldschmidt's . . . ideas *were ridiculed* for many years . . . [However, in the 1970s] Niles Eldredge, curator of Invertebrate Paleontology at the American Museum, was collaborating with Dr. Stephen Jay Gould of Harvard, and calling their new theory, aimed at explaining the GAPS *"punctuated equilibria."* [or PE]'[59]

Eldredge and Gould, like Goldschmidt, wanted to explain the *sudden, non-evolutionary,* appearances of new species, ***without*** needing to find the transitional fossils Darwin's which theory called for. They wanted to make sense of the fossils *as they actually are,* which was an admirable aim, without, however, being willing to admit that the lack of supporting fossil data *had, already, falsified* Darwin's theory, which was not admirable. Because, at that point, they stopped following the data, as science is supposed to do, not because the data was false, but because they didn't like the implications for Darwin's theory of doing so.

Unfortunately, Darwin had set the pattern for his future followers, because, as Shedinger writes in ***Darwin's Bluff,*** Darwin really disliked admitting that a scientific idea he was fond of, might be wrong. So, one of his regular tactics was to claim that the ***data was wrong***—not his latest theory. He, and too many of his supporters since he, have allowed this to *override* what should always have been their *scientific* priority, which was, and it is, to stay close to the data, to ***follow*** the evidence.

Art Battson observes that, *if we follow the data,* rather than trying to force it to fit with Darwin's ideas, what we see is that after the very geologically abrupt, the non-Darwinian, non-evolutionary appearances of new species, the *subsequent stability* of the species "suggests the existence of natural processes which *[actively] prevent* major change."[60]

So, as Battson reflects, Darwin *should* have inquired into the fact that the species *do not* gradually evolve into new forms, as he first thought, but remain very stable. He should have tried to *explain* this, rather than clinging to his own, pet theories, because he hoped that with his two key ideas of (i) *"descent with modification"* and (ii) *"natural selection,"* he might have discovered two new natural laws almost akin to the laws of gravity or electricity.

We, though, unlike Darwin, now know that *cells use very clever means* to ensure that they, precisely do NOT mutate in the way his theory requires. *Cells* work, very actively to PREVENT exactly the kinds of *random genetic changes* which Darwin's supporters say are the very sources of his imagined evolutionary process: microbe to snail, fish etc.[61] For more than 150 years Darwin's supporters have claimed that he successfully explained the origins of the species, but that there is *just not enough fossil evidence* to actually confirm his theory.

Battson writes, this "puts Darwin's theory in the curious position of explaining the data we *don't* have while ignoring the data we do,"[62] the tremendous STABILITY of the basic body plans and the fact that "NO TRANSITIONAL SPECIES [extinct or extant] showing evolution in progress between two stages HAS EVER BEEN FOUND."[63]

Darwin predicted slow, *non-jumping* change, but, as Latham writes,

> 'Check any textbook on the subject and you will see that SUDDEN APPEARANCE and DISCONTINUITY is the norm. The problem that Darwin so clearly saw, has not gone

away with time, and the FOSSIL RECORD seems now to REFUTE the entire basis of his theory of [very slow, very] gradual changes.'[64]

A mistake which many of us have made is to conflate the changes in the species, over time, which the fossils *do* show, with Darwin's theory, which they do *not* show. No, as Art Battson notes,

"Darwin's general theory of evolution may, in the final analysis, be little more than an *unwarranted extrapolation* from micro-evolution [minor variations, like finch beaks, or cleverly-selected-by-us dog varieties] *based more upon [materialist] philosophy* than upon fact. The problem is that Darwinism continues to distort natural science." [65]

Why does it distort science? Because Darwin's supporters—be they atheistic *or* religious—have been unwilling to admit that just because Darwin's theory was more scientific than trying to treat the Creation story in the Bible as if was a *scientific* description of the ***how*** of Creation, did not, of itself, make Darwin's theory true. Yet this is how so many of us now seem to think. If the Creation story in the Bible is not *literally* true, then Darwin's *blind-luck Creation Story* must be true.

In the first edition of *On the Origin of Species* Darwin wrote,

> 'In North America the black bear was seen by Hearne swimming for hours with widely open mouth, thus catching, like a whale, insects in the water. Even in so extreme a case as this, ... I can see no difficulty [*!?*] in a race of bears being rendered, by natural selection, more and more aquatic in their structure and habits, with larger and larger mouthS, till a creature was produced as monstrous as a whale.'[66]

Sadly for Darwin and his supporters, religious and non-religious, the fossils record no such fantastical transformations. New species appear *abruptly,* without plausible fossil ancestors, they *stably* endure, *without* changing into anything else, then, eventually, they disappear. Later, the next Life Wave of similar or somewhat different species appears in the *same, sudden, jumping way.* It too is stable. It too does *not* give birth to major new forms, and, in time, it ends. Darwin was aware that the *fossils did not bear out* his theory. He just hoped that, eventually, he would be proved right, and that the millions of fossilized species, gradually turning from one kind into another kind, which his theory needed, would be found. Unfortunately, for Darwin, they never have.

Still, acknowledging this fatal flaw in his proposals, Darwin asked,

> "Firstly, *why,* if species have descended from other species by insensibly fine gradations, do we not *everywhere* see innumerable transitional forms? [*'everywhere?!' Nowhere! do we see ANY* transitional forms, let alone 'innumerable.'] *Why* [he also asked] is not all nature *in confusion* instead of the species being, as we see them, well defined?" [67]

Darwin had a simple, linear idea. We know how a valley evolves, water shapes it. He wanted to apply this same kind of simple, linear thinking, which worked well enough in geomorphology, to *Life* itself.

But, *he had no data* that such a process *'might'* happen, let alone that it did. One thing, we can say, though, is that, like a lot of the things we make, "Many living things are connected by design." For example, many shared skeletal patterns and shared *instincts* and *soul* qualities.

In *The Naked Emperor: Darwinism Exposed,* Dr. Latham writes that by looking at the fossils of the earliest four-legged animals we can see a continuity between "the lobe-finned precursors" and the first land walkers. "They retain a sort of fish shape," he notes, but, the ABRUPT LEAP to land-walking is *unexplained,* and the early, four-legged, land-walkers are not in any way *"transitional"* in the way Darwin meant.

He continues, "We see too much that is suddenly new to call the first tetrapods [four legged land walkers] transitional. There is *no gradual evolution* here but we do see earlier forms being a sort of template for the creation of later forms. The appearance of tetrapods is just one of the *saltations* [L E A P S of form] that characterize the fossil record." [68]

Yet Darwin was insistent that evolution could *not* make L E A P S. It *had* to be gradual to be true—very slow change. Except, the *transitional or linking* fossils his theory needed are *never to be found,* never mind the "inconceivably great" numbers he called for (see cover).

Latham, "The transition to tetrapod is *sudden.* Darwinists will put this down to *inadequate fossil preservation* but this argument is now *wearing thin*. The challenge to the Darwinist is to SHOW US EVEN ONE CASE in the history of life where a macro [evolutionary] change has occurred smoothly with well-defined, gradually transitional intermediate forms. DARWIN WAITED AND WE STILL WAIT.'[69]

Geoffrey Simmons mentions a few more of the many Living things which cannot be explained or explained away by appeals to vast time in combination with [impossibly unrealistic] lucky chemical chance.

> 'Darwin wrote that whales came about as a result of *bears* going to sea. [Yet, no one has a clue] how blow holes came about…or how the internal lungs became connected up to these holes in a way that prevents drowning. Or, how a *massive communication center,* found in their heads, came about. Or, how the *ability to depressurize body segments* during deep dives evolved [due to random mutations]. …
>
> 'Whales are not the only misfit to smooth transitions … *kangaroos, woodpeckers, platypuses, giraffes, butterflies, octopuses, skunks, bombardier beetles, the red tide, dolphins, fireflies, tardigrades, sloths, and all microorganisms ...*
>
> 'It's true, natural selection happens in a variety of situations, but IT DOESN'T CHANGE ONE SPECIES INTO ANOTHER.' [70]

Antony Latham writes, "the challenge to the Darwinist is to SHOW US EVEN ONE CASE in the history of life where a macro [evolutionary] change has occurred…with well-defined, gradually transitional…forms. Darwin waited and WE STILL WAIT." [71]

Why has our current [2026] science culture still not, so far, openly accepted that Darwin's theory has been falsified? Because, as Lewontin put it, of its science-limiting commitment to **materialism**.

6 Darwin's Fatally Flawed Law

Darwin's idea was that Life *might* have started, by chance, in a 'warm little pond,' as a basic, single, Cell. Then, *might* such a basic Cell, like a Bacterium, VASTLY vary?, and VASTLY vary?, and VASTLY vary?, until slowly it evolved, due to millions of changes, into far more complex *multi-celled Beings* like *Worms—Snails—Spiders—or Fish?*

Might a two sex world 'evolve?' Male and female, with very different parts? Penises for some, ovaries for others? Although, dear reader, *please let us think!, just a little!, for ourselves!*: HOW could this happen, by pure c*hance?* Two *separate—yet totally in-sync—*evolutions, one for testicles, one for ovaries, due to *random* genetic mutations?[72] This idea, like so much else in Darwin's scheme doesn't make sense.

Might tiny microbes evolve into fish with backbones? [see cover].

Might water-breathing fish evolve into *air-breathing* amphibians?

Might amphibians evolve into the first dinosaurs?

Might land-running shrews evolve into sonar equipped bats?

Might land-living bears evolve into deep-sea diving, echo-locating, water-birthing, whales, with specialized, collapsible lungs, enabling them to dive to amazing depths, 10,000 feet!, at incredible pressures? [see cover]. How, though, did dear Darwin think that such VAST levels of change might occur? *He didn't know*. Really, his great theory was just an idea. So what was his idea? It was that all Life had descended, like a *branching tree,* from just one or two, *universally shared,* microbial ancestors, due to their *VASTLY varying,* over vast time.

He called his new law of *random variations* to *existing* organisms repeatedly arising, then multiplying, ***Natural Selection***. Yet all his so called *'new law'* was telling us was this: "That which survives, survives. *Why* does it survive? Because it is the 'fittest.' How do we know it's the 'fittest?' Because it survives." A circular argument. Although, *misleadingly,* Darwin called his book, ***On the Origin of Species,*** this was, ironically, the very point he did not answer. Nor was it a question which his new law of natural selection could answer: "How do Living things arise in the first place?" *Natural Selection cannot tell us.*

Darwin also referred to his idea as ***Descent with Modification,*** by which he meant that all living things had descended from just a few ancestral microbes—like algae or bacteria—versus the traditional view that the species had arisen separately and were, relatively, unchanging.

He, by contrast, asked, *"What if* organisms were almost *infinitely* plastic, and *not stable at all? What if random* changes occurred in an *existing* organism, like a microbe, and those changes were *beneficial and heritable?* If so, its microbial offspring ought to *do better* than the others of its own species, and it's descendants would predominate.

What if a *microbe* could vary so HUGELY, due to enough random mutations [millions?], as to, eventually, transform into a VASTLY LARGER Worm, Bat, Butterfly, Eagle, or Fish? [see covers].

What if larger animals, like Bears, undergoing *millions* of such random changes, gradually evolved into *Whales?* [see book's covers].

Those organisms, which had so varied, with *heritable traits* favoring their variant's survival, would leave *more offspring,* causing those traits to increase over the generations. For example, the more *whale-like* Bears would, slowly, do better, in the seas, than the *less* whale-like ones.

The early, *less spidery,* Spiders would not do as well as the later, *more spidery,* Spiders. The first rodents would not be very *bat-like,* nor would they be equipped with *sonar, more efficient* than any sonar we make. But, gradually, due to millions of random variations, the highly efficient, sonar-equipped Bats, we now know, would triumph.

This is why Darwin wrote that the *number of transitional species,* of bats, spiders, bears and whales, which should show this amazing process, must have been *"**inconceivably great,**"* if his hunch was true.

"Inconceivably great!" Let's imagine it. Countless 'trial' versions, that *failed.* Bears starting to become whale-like, some failing, some succeeding.

Rodents starting to become Bats, with *many* fossilized bat-failures along the way. Only, there's absolutely *no evidence* for them.

But leaving aside the [total] LACK of supporting evidence, once Darwin's theoretical universal ancestor microbe had begun to vary, it would, he argued, *vary* in different directions. For example, *might* a tiny microbe become a VASTLY LARGER Worm? [***How?!***] First offshoot. Another a hard shelled ammonite?! [***How?!***] Second offshoot. Another a vertebrate fish? [***How?!***] Each *new anatomical design* would be a new offshoot. These would be the first branches on his "tree of evolution."

But, imagine the *staggering level* of change needed for a tiny, single-cell bacterium to turn into a VASTLY LARGER FISH?, [see cover], or Spider?, or, eventually, Whale? All this, we are to believe, due to a totally random, *no-intelligence-needed process of random genetic mutations!* Not to mention that there's *no evidence* for any such process.

You may be wondering, dear reader, how Darwin's contemporary colleagues reacted to his ideas? Did they take them seriously?

No, far from it, mostly they did not. Despite not having our modern knowledge of the *truly mind-boggling complexities* of living things, there were plenty of scientists who were skeptical of his claims, which, they thought, were *fanciful,* unrealistic and *unsupported* by the data.

For example, St. G. Mivart, who published *On the Genesis of Species,* wherein *he doubted* that natural selection could *initiate* biological structures. S. J. Gould wrote of Mivart that,

"Darwin offered strong, if grudging, praise and took Mivart far more seriously than any other critic... Mivart gathered, and illustrated "with admirable art and force" (Darwin's words), all objections to the theory of natural selection. . . "a formidable array" (Darwin's words).

"Yet one particular theme, urged with special attention by Mivart, stood out as the centerpiece of his criticism. ... No other criticism seems so troubling, so obviously and evidently "right" (against a Darwinian claim that seems intuitively paradoxical and improbable). Namely, as Mivart referred to it: "The Incompetency of 'Natural Selection' to account for the Incipient Stages of Useful Structures." [73]

Mivart made the point that natural selection cannot *initiate* living things. Why not? Because it has *no reason* to select any chemicals in the first place, let alone the *very specific* combinations needed.

Darwin based his ideas partly on his knowledge of the changes that ***intelligent*** people, with ***intelligent*** aims, could bring to an ***existing*** species. People who *already* had something very cleverly put together, like Wolves, Pigeons, or Tulips to select from. Whereas, evolution by natural selection relies on *r*An*Dom* changes for all its alleged creativity.

Yet there is no reason for such an evolution even to *begin. HOW could it? HOW could Cells,* arise by chance? Nature 'choosing' the fittest variations to *existing* species, natural selection, *cannot explain* the very *start* of Life. Because, at that stage, there are no *fit or unfit* species to vary, or, to choose from. There are no *fit or unfit Beings to* survive.

Nature's selection, via the superior survival rates of the *fittest* variations, *can only refine*, to a (strictly) limited degree, what *already so cleverly exists*. It cannot generate the truly new. How can it?

The creation of any purposeful mechanism, *including Living things, needs mind and creativity.* But Darwinian evolution *has* no mind. Yet, this is modern science's ***most extraordinary claim*** backed up by *no evidence* at all, let alone *any extraordinary* evidence.

Life is just a long series of chemical flukes. It doesn't work, it doesn't make sense, it's illogical, but our *materialist* friends say they can't believe in miracles—apart from ones of BLIND CHANCE!

Yet, every living thing is a miracle, isn't it? Cells!, blood, DNA Codes, Hearts!, Lungs, Blood, Brains, Arms, Hands, Eyes, Minds!, Spiders! Bats, Bears, Whales, People! Created by PURE LUCK? *HOW?'*

The idea that *Life did* arise, and *did* become ever more sophisticated, by totally MINDLESS and DIRECTIONLESS processes, has never had a justifiable *scientific* basis. Yet, most of those who do so firmly believe that there cannot be any *intelligent or spiritually causative* components to reality—which ***just Are***—seem strangely unaware of just how much *utter non-sense* we need to believe in order to try to make their *atheistic* beliefs in blind luck which ***just is***—as a great creative force—seem true.

Yet, when we point out that their scheme is *not just unscientific,* but demonstrably impossible, they laugh at us, while insisting that *they* are the true defenders of *'science.'* Unfortunately, once Darwin's *no-intelligence-is-needed-to-create-living-things* ideas caught on, so many thinkers did not notice that their own newly minted beliefs in BLIND *chemical* LUCK, as a supremely creative force, were, *oh how ironic!,* a vastly weaker way of thinking than some of the old religious ideas which they wished to replace—with their newfound atheistic *faith.*

At least the traditional religions understood that it makes sense, overall, to hypothesize that *intelligent Life* emerges from a matrix, a *multi-dimensional Matrix,* which is, itself, *intelligent.* The idea that MERE LUCK can create anything remotely clever never had a *scientific* foundation. It was just a strange *new faith.* Thankfully, some scientists did question the *new faith.* They pointed out that natural selection *can only* account for the *survival,* or, at best, the minor variation, of what *already* exists. It can not account for its arrival in the first place.

In ***Darwinism: The Refutation of a Myth,*** Soren Lovtrup wrote, "some critics turned against Darwin's teachings for religious reasons, but they were a minority; most of his opponents ... argued on a completely scientific basis." Lovtrup continues, "the reasons for rejecting Darwin's proposal were many, but, first of all, that many [biological] innovations cannot possibly come into existence through the accumulation of many small steps... because incipient [early] and *intermediate stages are not advantageous.*'[74]

Natural selection cannot *initiate Life.* It can only favor *existing Living things.* For example, the finches born with slightly *larger beaks,* will do better in certain climatic conditions than those with smaller beaks. But natural selection has neither the ability—it's mindless—nor reason—it's mindless—to select or to *livingly* order the highly *complex* mixes of chemicals needed for any living thing to arise in the first place.

Darwin, though, as a materialist, had, due to his beliefs, to see Life as 'JUST CHEMICALS.' Materialism has, by its definition, to deny the classical hypothesis of *any intelligent* components to the creation, and to the *second!, by second!, maintenance* of *Living things, blood circulating!, hearts beating!, livers treating!, kidneys purifying!*

The Life and Soul Forces versus the **INSENTIENT, DEAD, MATTER** of **MATERIALIST** *belief.* Yet, as Michael Flannery writes,

> "By and large, the scientists of his day were not much impressed with Darwin's theory. John Herschel called natural selection *"the law of higgledy-piggledy,"* and William Whewell thought the theory consisted of *"speculations"* that were "QUITE UNPROVED BY FACTS," [emphasis added] so much so that he refused to put the book on the shelves of the Trinity College Library."
>
> "Rather, it was the reading elite of London that was captivated with Darwin's theory. "Freethinking" bohemians and assorted society trendsetters grabbed up copies of the *Origin* and later *Descent*. The secular [atheistic] creation myth they had long been looking for was finally in hand. ... It was a pitch easily made because it "sold" a product the intellectual elites had long been waiting for, a theory of life in which God was superfluous and irrelevant." [75]

Darwin's idea, "Nature selects the best variations," *cannot* build the functional from *nothing*. All it can do is favor *existing* species members best suited to the current conditions. The offspring of the bears who are, genetically, the *best swimmers,* will *not* become *whales,* but they will survive better than the offspring who are more genetically feeble.

Ironically, Darwin was aware that *no species* had ever, *nor yet has ever,* been bred, even by the most skillful breeders, using *careful, clever, artificial selection,* to vary beyond a certain point without damage. Yet, *ignoring* the inability of natural selection to originate anything clever in the first place, the *actual origin* of species, he argued that a random change in an *existing* plant or creature *might* arise, it *might* survive, and, via an *"inconceivably great"* [his words] number of transitional versions it *might,* eventually, lead to a new species, then to another new species, and another, and so on, for ever. Only, the fossils have never supported his speculations, microbe to snail… bear to whale.

In ***Darwin's Bluff, the Mystery of the Book Darwin Never Finished,*** Robert Shedinger reminds us of Darwin's own, self-excusing, words in *"On the Origin of Species,"* which he first published in 1859.

Darwin wrote, "My work is now nearly finished; but *as* it will take me two or three more years to complete it, and *as my health is far from strong,* I have been urged to publish… some BRIEF EXTRACTS from my manuscripts. This ABSTRACT, which I now publish, must necessarily be imperfect." For, *"I cannot give references* and authorities for my several statements…" Hold on, why not? Although if he was *bluffing* as to the real existence of convincing supporting evidence for his theory, from the fossils, this was a useful ploy.

So, playing for time, Darwin wrote, "I can here give only the general conclusions at which I have arrived, with a FEW FACTS in illustration, but which, I hope, in most cases will suffice." He continued,

"No one can feel more sensible than I of the necessity of *hereafter* publishing in detail, all the facts, with REFERENCES, on which my conclusions have been grounded; and I hope in a future work to do this.

"For, I am well aware that *scarcely a single point* is discussed in this volume on which FACTS cannot be adduced, often *apparently* leading to conclusions directly *opposite* to those at which I have arrived.

"A FAIR RESULT can be obtained only by *fully stating* and balancing the facts and arguments on *both* sides of each question; and this *cannot possibly* be here done."[76] Hold on, Darwin, why not?

Why did Darwin not "state the arguments on both sides?" Because it did not suit him? We must also question his word, *"apparently,"* because, if we do look at "both sides," a "fair result" argues, *not* for the descent of all *Living Beings* from a universal microbial ancestor, but for their *separate* origins and their *separate* descents. So, did Darwin, later, publish, "in detail," "all the FACTS, with references" on which his "conclusions were grounded?" No, he did not. Yet, in his later exchanges with other scientists, he repeatedly said that *"On the Origin of Species"* was but an introduction to his ideas, which he would later substantiate in a more authoritative work, and it would take him just "2 or 3 more years to complete it." Two or three more years is not long.

So his later work, backed up by far more FACTS, illustrations, and proper REFERENCES should have been out by 1862. Yet, despite his repeated promises to his scientific colleagues, it never came. Why not? Darwin lived on for twenty years, until 1882. His work was "nearly finished," but all we the public, and his fellow scientists, ever got to see were those few "brief extracts" from his allegedly much larger work.

Was it a *bluff?* This would not matter *if* the ideas Darwin first outlined in 1859, in *On the Origin of Species,* had, in time, been borne out by the *actual data*—as he hoped might happen—it *never did.*

But, ignoring the lack of data needed to support his theory—like fossils of *bears slowly becoming whales* or *shrews* slowly becoming *bats,* or of *un-spiders* gradually becoming *spiders*—Darwin just hoped that he might have hit upon one of reality's *key* organizing principles, a new natural law, and that only the *details* remained to be filled in.

For generations now, we have been assured that Darwin successfully explained Life's intriguing history. He did not. We need to hear the far more truthful: "It was a scientific idea of its steam-age, Victorian time, but it did not, in the end, prove to be correct."

Regrettably, Darwin's fans, then and now, have never accepted this. They have consistently refused to let the data speak, and to let Darwin go. Unfortunately, like Darwin himself, they tend to feel that his ideas are just 'too good' not to be true. So they say: '*They must be right!*'

I'm guessing you may be thinking, "That's not science, it's a *faith position—it's ideology."* You're right. The trouble is, there is so much more *politics* and *philosophy* going on in this contentious area than many of us realize. It is *naive* to think that all this is only about good, honest, science, and the detached pursuit of truth for its own sake.

For example, because so many of us now equate MATERIALISM and its associated philosophy of ATHEISM with reason, and we equate *spirituality* with unreason, we would rather pretend to ourselves that

Darwin's theory is correct, than risk letting *Soul and Spirit* back into our understandings of *Living Nature*—if we admit that his *bluff* has failed.

On the face of it, Darwin's theory was not illogical: First, there was a universal ancestor microbe, which *varied so* VASTLY as to become a vast Tree of Life. A simple, MECHANICAL, idea. One which *religious* believers could believe was *God or Spirit's* way of creating, while Darwin's *atheistic* fans could argue it was a wholly random process.

But the simplicity of Darwin's basic idea: tiny, microbe to VASTLY larger Worm, worm to Fish, rat to Bat, caterpillar to Butterfly, bear to Whale—*r*A*nd* *om* or *intelligently* driven—did not make it true.

Ironically, the real-world data of the fossils never obliged either his *atheistic* fans or his *spiritually* inclined followers. But Darwin was very reluctant to admit to the failure of his theory. This is why his fans, like the unhappy sisters in *Cinderella,* continue, to this day, to try to force the Darwin-refuting data of the real-world fossils, embryology and genetics onto the *hapless 'feet'* of his repeatedly disconfirmed, ideas.

What we *do* now know, but Darwin's supporters still do not fully concede, is that the superior survival rates of the fittest members of an existing species *can only refine* what ***already*** exists. *It cannot explain* the *arising* of *Cells, or Blood, or Livers, or Lungs, or Bones, or Brains, or Microbes or Worms or Spiders or Fish or Bears or Whales or People* in the first place. *HOW can it?* It can't. It should be obvious.

The mindless nature of materialist belief *cannot* create *anything remotely clever. HOW can it?* Natural selection *cannot* create *Cells. How can it?* Natural selection *cannot* tell us *how Life begins.* Natural selection *cannot* explain Life's DNA CODES.

Natural selection or NS *cannot* create CODE. Codes require *intelligence* to exist. They *cannot* arise *by chance.* NS *cannot* account for *intelligence and purpose* in nature. NS *cannot* explain the *sudden appearances* of the many weird *and* wonderful animals of the Cambrian Explosion. So called Natural Selection *cannot* explain *Mind or Thought.* Natural selection *cannot* explain *sentiency.* Natural selection *cannot* explain *goals* in minds. It has *no mind!*

Natural selection *cannot* transform an existing species into a different one, and, there is *no evidence* that it has ever done so. Yet, ***aimless, unguided,*** natural selection was Darwin's ***core idea***. The word 'selection' *implies choice* and *purpose,* but the *unintelligent* Nature of Darwin's belief system cannot truly *'choose'* anything, because it *has* no aim or mind. Are these not all true statements? But, because they conflict with the MATERIALIST belief system they are ignored.

In ***Natural Selection or Natural Creation****? – A Complete New Theory of Evolution,* John Davidson discusses the Hindu-Yogic notions of creation as an *Eons Long Creation in Intelligent Evolution* or ELCIE.

Davidson outlines just some of the *many fatal flaws* in Darwin's materialist and atheistic scheme. He asks,

> "What happens, then, if one attempts to understand this vision of nature's wholeness in terms of conventional [Darwinian] evolution theory...? One is immediately led to wonder ***how*** such wholeness could have arisen from linear and random processes? … "Let us take a specific example: the ***evolution of a spider***. The question is: *how* could a spider have evolved by random mutation?
>
> "A spider exhibits a host of integrated biochemical, … anatomical, behavioral and instinctive adaptations. It is able to manufacture ***perfect silk;*** to spin and weave it into perfectly organized webs and traps; to know how to walk in a sticky web without getting entangled; *to know* how to wrap up a fly, but to identify and leave a wasp alone; to … [wait] patiently at the edge or center of the web; to paralyze prey with ***so perfect a toxin*** that it does not then poison itself when it sucks out the body juices with specialized mouth-parts – and so much else besides . . .
>
> "So, musing upon how a spider could have evolved so many integrated skills and biological systems from some unspiderly ancestor, ***are we to suppose*** that the earliest spiders wove rather poorly structured webs out of inferior silk; got

stuck occasionally in their own webs; could not distinguish between dangerous hornets and houseflies; *used poor toxins* which only partially killed or paralyzed their prey while later giving them a ***tummy ache*** or worse; and that they were ill adapted for actually eating and digesting their prey and so they only relied partially upon their spiderly ways for making a living, anyway? In short, that they were pretty much of a mess, still needing considerable time, not to mention *several million "chance" mutations* [which are rare events and cells actively protect against], to get their act together?"

Davidson makes more interestingly valid points. He writes,

"Since such evolution is still supposed to be in progress, ***why are there no semi-evolved creatures living*** NOW? Why are there *none found* in the fossil record? In fact, 300 million-year-old spiders with web-weaving spinnerets have been found in the fossil record, ***yet nowhere has any fossilized species been found which is*** ON THE WAY to becoming a spider.

"It's spiders or something else! All creatures, not just spiders, are well structured wholes...Evolution by chance and by miniature steps would pull the creature in a host of different directions... For one cannot attribute any *goals* [progression] to a materialistic view of evolution.

"Nature cannot be assumed to know where she's going in such a philosophy of chance and chaos. And in the absence of any consciousness, vision or volition, how could such perfect designs as we see around us have come into existence?

"Yet somehow we are asked to believe that living creatures, who most clearly possess volition, desires, goals, *MIND* and *CONSCIOUSNESS* have arisen by random processes from DUST and WATER which exhibit none of these. MATERIALISTIC evolutionary theory is forced to assume that there is *no apparent purpose to life* and that natural selection is not selecting for some future, *intelligently* conceived goal; it is a selection only conferring advantages in the current moment.

"How then could some radically new species design [microbe, worm, snail, spider, fish, bird, feline, canine, ape, human] emerge ***purely by chance,*** and over an immense span of time, yet exhibit such a multitude of integrated characteristics?" [77]

'In 1998, molecular biologist Michael Denton wrote ***Nature's Destiny: How the Laws of Biology Reveal Purpose in the Universe.*** He wrote,

'The ***design of living systems,*** from an organismic level right down to the level of an individual protein, ***is so integrated*** that most attempts to engineer even a relatively minor functional change are bound to necessitate a host of subtle compensatory changes. It is ***hard to envisage*** a reality less amenable to Darwinian change via a succession of independent UNDIRECTED mutations altering one component of the organism at a time.' [78]

This is crucial, *millions* of highly integrated biochemical changes would be needed for darwinian-style evolution to take place. Yet random genetic mutations—which all cells so very cleverly and so very actively protect against—are rare events.

Or would we rather continue with Darwin's failed ideas just because we dislike the alternatives?

That there is, after all, *Purpose and Meaning to Life?*

That Nature, actually, is *intelligent, stunningly, staggeringly so?*

That we really are *Souls,* whose lives here, are but brief sojourns?

That Life after death is real, and the *immortality* of the *Soul* is real?

Why would we dislike the *Spiritualist* alternatives to ***materialism***?

Why would we not be glad about these things?'

7 Darwin's Imaginary Tree

The fossils never supported Darwin's idea that a *tiny* microbe might gradually change into a VASTLY larger Life form by, as Darwin put it,

> "differences not greater than we see between the natural and domestic varieties of the same species at the present day. ... [where] the number of *intermediate* and *transitional links,* between all living and extinct species, must have been INCONCEIVABLY GREAT. But assuredly, IF THIS THEORY BE TRUE, such [inconceivably vast numbers of transitional species must] have lived upon the earth."[79]

If it be true! We need, also, to forget about "inconceivably great." Because *no* convincing transitional fossils have ever been found, let alone vast numbers of them. In ***Darwin's House of Cards: A Journalist's Odyssey Through the Myths of Evolution,*** Tom Bethell recalls how he once toured the Natural History Museum in London with its then senior fossil expert, Colin Patterson. Patterson told Bethell "that he was looking for cases where Darwin's [imagined] shared ancestor of two later species was identified in the diagrams on display."

If they had once existed, they would be found at the branching points on Darwin's *hypothetical tree* of life. But, as Patterson showed Bethell, Darwin's *hypothetical branching points* on his hypothetical tree of evolution *were all non-hypothetically empty!*

Darwin's shared ancestors were never to be found so they could not be displayed on the diagrams in the specimen cases in the museum.

Patterson told Bethell that, "as far as he could see," the nodes, the places where Darwin's common ancestors ought, in his theory, to be, "are *always empty* in the tree of life," and he doubted they'd ever be filled. Tom Bethel then explores *"Darwin's Mistake."* He writes,

> 'We all know that small differences are observed between the generations. [For example finch beak variations and variably hued moths of the same species.] ... Darwin's mistake was to assume that these [minor] differences *somehow accumulate* over the millennia, so that one species eventually transforms itself into another.
>
> 'WITHOUT EVIDENCE Darwin's supporters today *still* [believe this]. ... But such a transformation HAS NEVER BEEN OBSERVED [either in living history or in the fossil records]. *No species* [fossil or living] has ever been seen [or known] to evolve into another.' [80]

Some refer to these minor generational differences as micro-evolution. But it is a *misleading* term. Because what they are calling micro-evolution is *directionless* and reversible. For example, the proportions of finches with different sized beaks varies from year to year, as climatic conditions change. But, these variations are not cumulative or progressive in the way Darwin had in mind, for his theory of long term, large-scale evolution: microbe to snail, fish or spider etc.

In an article, titled 'Demolishing Darwin's Tree,' *Evolution News (EN)* comments on a paper by researchers who are, "ready to provide a more *"expansive"* view of evolution" that replaces Darwin's tree of evolution, which claims to show the shared descent of the species, with a *"network"* of life. Why? Because "genetic data are not always tree-like."' However, *Evolution News & Science Today* asks, "If they replace the tree with a complex set of inter-connections [or web of life], what happens to" Darwin's core "notion of universal common descent?"

Evolution News quotes from the paper,

> "However, the more we learn about genomes the *less* tree-like we find their evolutionary history to be . . ."

'*Evolution News*: "Interesting: if "evolutionary history" is *not tree-like,* does [Darwin's key idea of] universal common ancestry [by descent from microbes into a vast, branching, tree of life] still hold? …

Calling it "historic," the authors recognize they are proposing a huge "shift away from strict tree-thinking to a more expansive view of what is possible in the development of genes, genomes, and organisms through time." *Evolution News* comments on this. It says: "They use "development ... through time" as a synonym for evolution. But what kind of evolution? If it is *not* tree-like, what is it? In a network… [there is] no clear progression from simple to complex." [81]

The *steady accumulation* of changes, leading, over time, to an amazing *progression,* are key to Darwin's ideas. VAST changes, from tiny microbes, like algae or bacteria to mighty beings like Oaks and Sequoia, Bears and Whales. A highly speculative, and unproven process, to which Darwin's imaginary tree is key. Yet, Michael Behe points out, *"**Darwin's "tree of life"** – where a single,* primordial cell gave rise to all subsequent organisms – *is **dead**.* The DNA sequence data *cannot* be made to fit with the idea." [82] It is ironic, too, that the *fossils,* so key to Darwin's ideas, *never* supported *him.*

Rather, the fossils show us that when new species arose, they did so *abruptly,* in a ***non-evolutionary*** pattern of ***sudden emergences,*** as in the Cambrian Explosion. "In any local area, a species does *not* arise gradually, by the steady transformation of its ancestors; it appears *all at once and fully formed."*[83] So wrote S. J. Gould. Earlier than Gould, another world class fossil expert, George Gaylord Simpson, stated,

> 'This is true of all thirty-two orders of mammals . . . In most cases the break is so sharp and the G A P so large [between Darwin's hypothetical common ancestors and their descendants] that the origin of the order is SPECULATIVE and MUCH DISPUTED…'
>
> 'This regular *absence* of transitional forms is not confined to mammals, but … is true of almost all classes of animals, both vertebrate and invertebrate … [It is true] of the major animal phyla [basic animal designs, like *Mollusca,* snails, *Nematoda,* worms, *Arthropoda,* spiders, *Insecta,* ants, bees, beetles, butterflies. *Crustacea,* shrimps, *Chordata,* fish,

amphibians, reptiles, birds, mammals, humans], and it is apparently also true of analogous categories of plants.' [84]

Simpson, strangely, then ignored the *Darwin-negating* implications of his own research. He wrote: "Man is the result of a *purposeless* and natural process that did not have him in mind." How did he *know?*

He didn't know. His confidence was *not scientifically* justified. It was just another statement of the *materialist* faith: particles-to-people, microbes-to-men, by PURE CHANCE. The truth, though, is that there is no good fossil data to support Darwin's idea of a universally shared descent—microbe-to-fish-to-frog-to-dinosaur-to-ape-to-man.

This is one of the odd things about this debate. When, after my own time as a Darwin fan, I began to doubt that *any* entirely *non-spiritual* theory of life, including his, could, *honestly,* explain the complexities of Life, I did not realize just how much the *actual facts* of natural history, including the fossils, did not bear out his ideas at all! I know it may be hard to believe, given all the pro-Darwin claims, but it's true.

Luther Sunderland noted, in *Darwin's Enigma,* that, ironically, the actual, real-world fossils "presented the *strongest single evidence against* his theory… because they *did not reveal any evidence* of a gradual and continuous evolution of life from a common ancestor, the [very] *proof he needed* to support his theory."

Darwin, however, "ignored the problem and found comfort in the *faith* that future explorations would… ultimately prove his theory correct." Now after decades of very thorough exploration, "the picture is *infinitely more vivid* and complete than it was in 1859."[85] So, Sunderland asks, do the "fossils reveal a *continuous progression* connecting all organisms to a common ancestor?"

No, despite all the searching, the fossils, "the only direct scientific evidences relating to the history of life, *still do not provide* any [of the fossil] evidence for which Darwin so fervently longed. The GAPS between major groups of organisms have been growing even wider and more undeniable. They can no longer be ignored or rationalized away with appeals to the imperfection of the fossil record."[86]

But *whence* this strange, *unscientific* passion to prove Darwin right, despite the data showing he was not right? Because this area of science is far more about a battle of ideologies, than it is about useful, everyday, *practical* science: it is about the battle between the **MATERIALIST** (naturalist) and `spiritualist` *(super-naturalist)* worldviews.

If it was purely about ***practical*** science, like whether a plane will safely fly, or about the efficacy and safety of some new drugs, the results would be accepted: "Darwin's plane does not fly. In fact, despite repeated revamps, it has never flown, and his drugs are not safe!"

But because, in this field, no one is going to die if we are wrong, we can persist, almost indefinitely, with *theories that don't work,* claim we are being scientific, when we are *not* being scientific, we are being **MATERIALISTIC**. The thing is if we have decided, from the off, that **MATERIALISM** is true, we are not going to give way to the *spiritualists* without a fight, are we? Even if, we are wrong and they are right.

No, what the magical fossils *actually* show us is the mysteriously SUDDEN appearance of new species, *not* Darwin's predictions. This is why trying to force the unhelpful fossil data, and the unhelpful data from embryology, genetics and molecular biology, to fit Darwin's theory, has been such a waste of time. I just hope that gradually more of us will start to *see* what, before, we did not see. That Darwin's ideas were not borne out by the *data,* and they *are,* now, more like a *faith.*

That, like the *Emperor's New Clothes,* we need to call Darwin out. Because, there is *no evidence* that microbes slowly turned into vastly LARGER WORMS, SPIDERS, or FISH. There is *no evidence* that Fish gradually transformed into Frogs—Fish-frish-frosh-frogh-Frog.

There is *no evidence* that Dinosaurs *gradually* turned into Birds. There is *no evidence* that land mammals, like Bears, *slowly* turned into sea going, echo-locating *Whales!* There is *no evidence* that land-running Shrews, gradually turned into sonar-guided Bats.

No, what the fossils ***actually show us*** is that new species appeared in a very *non-Darwinian,* and very *non-evolutionary* way: *abruptly.* Then, they *stably* lasted, *without* changing into anything else, before, in time, disappearing into extinction. Darwin *knew* that the fossils did not support his theory. He just hoped that, *eventually,* he'd be proved right.

However, as Schedinger says in *Darwin's Bluff,* he was far too attached to his theory to let it go, because of its potential to give him everlasting scientific glory, if he really had unravelled the mysteries of the history of Life? Did his ego get in the way of the data? Yet, admitting at least some of the problems with his ideas, Darwin wrote,

> "LONG before having arrived at this part of my work, a crowd of difficulties will have occurred to the reader [Yes!]. Some of them are so grave that to this day I can never reflect on them without being staggered [Yes!]; but, to the best of my judgment, the greater number are only apparent [No.], and those that are real are not, I think, fatal to my theory.' [This was *delusional*. Because so much is fatal to his theory.]
>
> "These difficulties and objections may be classed under the following heads: Firstly, *why,* if species have descended from other species by insensibly fine gradations, do we not <u>*everywhere*</u> see innumerable transitional forms? [<u>*everywhere*</u>*? nowhere!*] Why is not all nature in confusion instead of the species being, as we see them, well defined [and in separate kinds]? [*Correct.*] Secondly, is it possible that an animal having, for instance, the *structure and habits of a bat,* could have been formed by the modification of some animal [like a non-flying rodent] with wholly different habits?"

No, because not only is it *theoretically* unlikely, there's *no evidence* for any such process. So, if there is no data to show that Evolution, as Darwin described it, COULD happen, and there is no evidence to show that it DID happen, it is fair to conclude that it DID NOT happen!

Clever human breeders have never been able to *artificially* create a new species—so the idea that, rare, random genetic mutations, which are generally harmful, *not* beneficial, could turn some land-running rodent, into a bat, *by pure chance,* is not scientifically plausible.

Darwin, expressing his own—highly justified—doubts, wrote,

> "*Can we believe* that natural selection could produce, on the one hand, organs of trifling importance, such as the tail of a giraffe, which serves as a fly-flapper, and, on the other hand, organs of such wonderful structure, as the eye, of which we hardly as yet fully understand the inimitable perfection?"

No, we can't, and, importantly, there is *no evidence* that the giraffe's tail was produced by natural selection, let alone the eye.

> "Thirdly, can instincts be acquired and modified through natural selection? What shall we say to so marvelous an instinct as that which leads the bee to make cells, which have practically anticipated the discoveries of profound mathematicians?'[87] [Well, our materialist friends do assert that Bacteria arose by chance, Beehive making arose by chance. Spiders webs arose by chance. Eggs arose by chance...]

What a relief, at least, to see that Darwin had his own doubts. But, if we are forced by our philosophy, as he was, to deny any part to the *super* natural in the creation and maintenance of living things, *this second!, and every second!, now!, now!, and now!, our astonishing Hearts Beating, our amazing Blood Circulating, our Lungs, Livers and Kidneys ceaselessly working,* we *do have* to attribute *impossibly unrealistic* abilities to BLIND CHEMICAL LUCK to create and to "improve" on, "organs of such wonderful structure, as the eye, of which we hardly as yet fully understand the inimitable perfection." (Darwin.)

This *was* Darwin's dilemma, as it is for all our materialist friends. Keep to their *atheistic* belief system and argue for the *scientifically* impossible, or, with a sigh of relief admit to its failure, and be free of its *unscientific* insistence on blind luck explanations for everything.

What a relief it will be when our sciences are freed, at last, from Darwin's *impossible, no-intelligence-allowed* paradigm. What a relief when we admit, once again, as a matter not just of commonsense, but of logical, sober, science, to the *amazing!, Living!, intelligence!* which keeps our own *stunningly!, stupendously!, complex* bodies *working,*

24/7, lungs breathing!, hearts beating!, right now!, this amazing moment!, and every amazing moment!, now... and now... and now...

Ah, but we do *so dislike* to admit that we don't *know!* So, some of us, either take the old religious creation stories too literally, or, we resort to scientific-sounding *'Just So'* stories, like Darwin's, to try to explain things. Then, when that, also, does not work, we do so hate to admit it!

Too many of us do now seem to think that our only choices are between either (a) 'superstition,' which we equate with all that is *religious* and *spiritual,* or (b) 'science' and 'reason' which we equate with materialism, atheism, and Darwinism.

Whether an *intelligently driven* creation, versus the Darwinian-Materialist proposal of a mindless, blind-chance driven 'creation,' is the truth, is a valid scientific question, which we can try to answer.

How accurate the world's *spiritual* traditions, like Hinduism, Christianity, and Islam are in their depictions of the *super* natural and the *spiritual* are separate questions: Questions like whether God is *One?* —Islam, *Three?*—Christianity, or *One* (Brahman), *Three* (Brahma, Vishnu, Shiva), *and Many?* (other gods)—as in Hinduism, or *no 'god,'* in the traditional sense, at all?—as in Buddhism and Taoism.

Or fascinating questions like: "What happens *when we die?* Is there any truth to the old religious depictions of the afterlife, with horrible hells, that only sadistic gods [made in our image?] could *devise?,* or boringly *pious heavens?* Is there any truth to the notion of human to human *reincarnation,* after a spell in *spiritual realms?*[88]

Is *God or Spirit perfect,* as some say, or is *Spirit itself evolving?* Creation, even if it is *intelligent,* hardly seems *'perfect'* from our, ordinary, human perspectives, here on this so-harsh-at-times planet.

Personally, I think that while the traditional religions have the *basic* idea of the *spiritual* right, they are all, probably, inaccurate to some degree. This is where the plentiful *modern psychic research* into the *super* natural[89] comes in. Regrettably, we do not hear much about this *modern research*—which provides such convincing and, frankly, overwhelming, *scientific-spiritualist evidence* for the *spiritual*—because it *contradicts* materialism, and some of the old religious dogmas too, so the mainstream religions also tend ignore it.[90]

Another contradiction in the modern *materialist* approach, which so dominates science today, is, how can something, which, in its view, has no intelligence or purpose to it, be *"improved,"* as Darwin put it?

Because *'improvement'* always implies *existing* function, doesn't it?

Yet, as soon as a *Cell, DNA, Blood, Heart, Lung, Liver, Kidney, Ear, or Eye,* functions, it *must* have purpose to it—and reason behind it—otherwise it would not *work, or function,* at all. So without *purposeful forces* at work in *Living* things, nothing would ever work. Yet *purpose* always points to *intelligence.* But purpose and intelligence always have a source, *which implies subjective Being or Beings with purposes,* capable of *imagining* (creativity), that which is *desired* (love), and *actioning* it (creating): which is *intelligence—love—will*—in action.

Intelligent mechanisms require *intelligence(s)* to imagine *[mind-head],* to desire *[love-heart],* and to create them *[will-belly-limbs].*

The idea that Nature's amazing expressions are not *teleological* or purposeful, as our current science claims, so clearly, untrue. Ah, dear reader, so much is now so topsy-turvy, back-to-front and up-side down.

Lies have become 'truth' and truths lies. I used to think that *science* was the most rational and logical of activities. But when it comes to some of the most important things we can think about, including origins, it [currently, in 2026, still] feeds us nonsense.

Perhaps our amazing scientific and technical successes, in recent times, have, kind of, made us lose our minds. Ego? Vanity? Hubris? "Those whom the gods would destroy they first make mad…?"

Another of the downsides to Darwin's way of thinking, I noticed, when I began to pay more attention, was that it had nothing useful to say about the *inner Being, the Mind, the Soul, the Subjectivity and the Sentiency* in living things. Nothing about the *'hard'* problems of *S E N T I E N C Y and C O N S C I O U S N E S S.* Darwin's theory relates purely to the FORM or 'tails' sides of nature's holistic coin, to the

MATTER and MECHANISM aspects only, the beautiful BODY of Nature only. It is solely about Nature's OUTER forms. It has nothing useful to say about her *inner or 'Heads' sides, her Subjective* elements. Nothing about *Soul* or *Intelligence* in Nature. This was odd. Because, Darwin's `ideas` emerged, from his *intelligent* `Mind,` and his *curious Soul* trying, very *purposefully,* to make sense of nature, without, he argued, there needing to be any purpose, or *Soul,* or intelligence in nature.

Like many thinkers, philosophers and scientists of more recent times, he left *himself,* his *intelligence,* and his *purposes,* out of nature.

He looked at nature as if she was all OBJECT and FORM, all beautiful 'BODY,' but no *Soul,* no *Spirit,* no *inner* aspects—all very OBJECTIFYING, DEAD, LINEAR, MECHANICAL. But this, it is sad to say, is what MATERIALISM is. There can be no *Soul or Spirit or Anima* in its DEAD-EYED VIEW. For materialism there can only be MINDLESS DEAD MATTER which only *seems* to be *ensouled and alive.*

This, is our current mental mess, our *Faustian* obsession with the purely PHYSICAL aspects of Nature, seeing her as *all beautiful* BODY, but no *Spirit or Soul,* while it has brought us many material benefits, many luxuries and indulgences, has, tragically, cut us off from the *true Life and true Soul* in all things, from the *deeper meanings* of all things.

Thankfully, some of the world's earlier peoples still remember—for the rest of us. For them, *all of reality,* from rocks to oceans, plants to animals, planets to stars, even the vast, gyring galaxies, have *inner, Subjective, Soul, and Spirit* aspects.[91] For they, the entire universe, at all levels and dimensions, atom-to-mineral-to-cell-to-plant-to-people-to-planet-to-star-to-galaxy, all the way up and all the way down, all the way out and all the way in, is an expression of `Living, Creating Spirit`, it's *not* all just DULL, DEAD, MATTER, the *panpsychic* view.

Interestingly, the *Yogi-Adepts* of India who are old-earth creationists in their spiritual vision, maintain that, over the eons, Life on Earth has waxed and waned across vast Life cycles, called *Yugas,* some of which *Life Waves* may correspond to the mysteriously sudden emergences of the new species, as shown to us by the fossil records—which Darwin ignored.[92] According to the yogis, *Life's* history is not just an unfoldment of OUTER [mechanical] FORMS but a meaningful unfoldment of *sentient interiorities,* of diverse *Soul* states for the different kinds of beings. For example, the fascinating world of the dinosaurs had its *own, mysterious, Soul feel . . .*

Even now, we still can sense it... the amazing dinosaurs... It mysteriously arose, it served its mysterious purpose, and it went.

Of course, the *true nature* of Nature, of *What Is,* including of we, and of the amazing world around us, plants and animals, planets and stars, lies at the heart of all our debates about Life's ultimate origins.

While a purely physicalistic approach can be used to investigate what a *Star, Cell, Heart, Car or TV* is made of (what Aristotle referred to as its material cause) and how it works (efficient cause) inquiries which our sciences are now very good at, it does not tell us *how* or *why* they arose in the first place—about the ultimate causes of things which Aristotle referred to as its formal (shaping) and final (purpose) causes.

Our science has, for several centuries now, focused only on Aristotle's *'made of'* and *'how it works'* causes, while ignoring his other two causes, the ideas, intelligence, and purposes, behind nature's creations. The *unrealistic* conceit of our current science is that *nothing* in Nature *has* an intelligent cause. No, it all arises by pure chance.

Which, if true, means that nothing in Nature has any *teleology* or purpose either—which means that our Cells work for no reason, our Hearts beat for no reason, our Lungs work for no reason, because it's all just 'ACCIDENTS OF CHEMISTRY' isn't it?

If you are starting to think that this is an absurd way of looking at the world, I agree with you. Yet so many of our scientists, and so many of our other thinkers, currently profess to *materialism.* Why, though? Well, it's probably partly because, as Chris Carter says,

"In our modern world, *science and scientists* hold a great deal of prestige, and few people want to be thought of as unscientific. ...

"[So] If to be scientific is good and unscientific bad, and if the term "scientific" is thought to be synonymous with the term "materialistic," then any talk of *disembodied minds or spirits or [of intelligent causation]* is anti-materialist, unscientific, and therefore bad. [93]

"The long-standing *confusion* of materialism with science is what largely accounts for the persistent social taboo responsible for the ignorance and dismissal of the substantial amount of evidence that proves materialism false." [94]

Ah yes. So many of us do now seem to think that our amazing modern capacities to work out how what living things are ***made of*** and ***how they work,*** Aristotle's first two causes, justifies us to say that they have *no intelligent* causes, no, what Aristotle called, formal and final causes, despite them being far more complex than anything we make.

Yet, does it not make much more sense to see Nature as *Intelligent and Purposeful?* Does that not match the data, overall, so much better? To realize that *Intelligence, and Soul and Spirit, inhere* in Nature?

Some refer to this way of seeing nature as ***pan-psychism,*** where *Self-Existing Being, or Spirit, which* ***just Is,*** not only *gives rise* to all things, but is ***within*** all things, *enlivening, and pervading,* all things. According to this understanding, *Spirit-Consciousness-Being* is not only *transcendent* to creation, but *immanent to it, now, now, and now, 24/7.*

Our world's earlier peoples understood this. They experienced the natural worlds as *en-Spirited and en-Souled.* For them all things had *inner, Subjective, Soul, and Spirit dimensions*. Unfortunately, though, despite all our modern cleverness, this *inborn,* intuitive sense of the ***inner Livingness,*** *and* ***inner Soulfulness*** of all things, has, for now, given way to the dead-eyed-god we call materialism.

This is why, especially after Darwin, scientists began to assume that, either, *Soul and Spirit* do not exist at all, or, if they *do,* it will, still, be possible to explain the *Living Worlds as if* they do not. They say:

"Regardless of whether *Life!* has intelligent sources, we can explain it in purely *'natural'* terms. It may *look* highly intelligent or designed, but, in the right conditions, it could *fall together* by pure luck. *"No intelligence was ever or is ever needed!"* But, just as this approach won't work for a Car, PC or TV, it's unlikely to work for nature's organisms, all far more complex than anything we make. Equally, if we have *immaterial Souls,* which continue, after death, into non-physical dimensions, are we not part of an over-arching *Spiritual Nature* which is not only *pervasive* but *multi-dimensional?*

This is why I prefer the terms MATERIALIST and MATERIALISM, over *naturalist* and *naturalism,* to describe the totally materialistic and mechanical way of thinking which is currently considered scientific.

'Naturalism' is ambiguous, because, while some see Nature as physical only, others see her as *multi-dimensional,* with *spiritual* elements. *"What* is the *true* nature of Nature?" is the very question.

MATERIALISM, right or wrong, is clear: the idea that only physical stuff, like CARBON OR H2O, is real. *Life and Mind* are *not* principles in themselves. No, our *Living Awarenesses, our Curiosity, our Creativity, our Love and Joy,* are all just ACCIDENTAL BY-PRODUCTS of a purely physical STUFF which is, itself, MINDLESS and DEAD.

For a while, as a teenager, after a first learned about Darwin and his ideas, I also began to think in this sterile, *Soul*-denying way.

But to conclude, as I did back then, that, just because the world's traditional religious creation stories are not *scientific* descriptions of Life's origins, our existence is just a long series of impossibly unlikely chemical coincidences seems unreal to me now.

No, there is clearly far more at work in *Nature,* than we can explain in purely physical terms. Call it *Soul,* call it *Spirit,* call it the *Life Forces,* call it *Living Being,* the names don't matter.

Our conclusions, though, do matter.

Do we, our amazing plant brothers and sisters, and our magical animal friends, live in a *miraculous, Living, Cosmos of Intelligence and Meaning,* or in the MINDLESS, DEAD, UNIVERSE of modern materialistic belief?

Which makes more sense?

Which is the more *intellectually fulfilling,* overall?

8 Why So Much Philosophy?

For a book on Evolution, why so much philosophy? Because our *current* origins science *is* so much more about competing *philosophies* and competing *worldviews,* materialist vs *Spiritualist*, naturalist vs *super-naturalist* than it is about *practical,* down to earth, science.

I didn't realize this for a while. I kept asking myself, "Why don't we see through all this, when so much doesn't make sense in our current ideas about origins—like Darwin's warm pond idea or his *Just So* story that Bears could, gradually, transform into Whales?" Why didn't more of our scientists call out: "Darwin's *Emperor* is Naked!" Then it hit me. It was because neither did they *wish* to, nor did they *need* to. Why?

Because, as Chris Carter says, scientists now "hold a great deal of prestige,"[95] some of which flows from their claim to have solved the greatest mystery of all, the very origins of Life, conveniently forgetting, of course, that *Darwin did not, actually,* explain the true origins of the species at all! His speculation, that Life *might have* begun by chance in a warm little pond, and evolved from there, was just that—speculation.

But, since 1859, when he first published, there has been a huge shift in the intellectual balance of power, between those who claim to speak for the *spiritual,* and those who try to make *physical-scientific* sense of the world. In the western world, especially, it was a rout. After Darwin did, so firmly, purport to explain Life's history, there was a dramatic, collapse in *religious* and *spiritual* belief, which is why, especially in the west, so many churches and other religious buildings now stand empty.

Why, therefore, would scientists wish to risk their wonderful newfound prestige by questioning (a) their pet theories on origins in general and (b) their *even more* radical claim that we no longer need *religion or spiritualism* to tell us anything about the origins, let alone the *deeper meanings of Life, the Universe and Everything.* Because, so many of them now, like Darwin in his day, feel that there aren't any!

Materialism, they say, is the better view and *Spiritualism,* the classical belief in *Intelligent Creation* and in *Intelligent Design,* is false.

What better way to destroy your intellectual opponents, then, than to argue that the area in which they claim to be experts, either (a) does not exist at all or (b) that, *even if* it does exist, it, either, (i) lies in some totally separate realm, with no bearing on the natural *physical* worlds, or (ii) that, even if it *does* have any bearing, we can, nonetheless, explain all things in *Living Nature*—no matter how complex—*as if* it does not.

This is the approach in modern science known as methodological materialism or methodological naturalism (MM/MN). It means that if either (a) or (b) is true, it makes anything to do with *religion,* or the *supernatural,* or the *spiritual,* irrelevant to the modern *scientific* process: "Philosophers, Priests and Theologians!, Listen up!, your views are no longer needed, when it comes to making sense of *Life!"*

It wasn't always like this. In ancient times, science and religion went hand in hand. The priests were often also the scientists of their time, in charge of the calendars, with accurate knowledge of the seasons, the movements of the stars, eclipses, and so on. But times changed.

This is why we cannot now discuss these deep subjects without considering the *philosophical* background, especially as so much more is at stake here, than the kind of *practical science* which will ensure that a plane will fly, a cell phone will work, or a new drug will be safe.

In ancient times, religion and science co-operated. But, more recently, the success of the natural philosophers, now called *'scientists,'* from the latin *'scire,' to know,* in unraveling the mysteries of the natural physical worlds, led to the scientists increasing confidence, and, in time, to what was, in effect, an epoch-making intellectual coup d'état, where they seemingly overtook, and overthrew, their one-time mental colleagues, the priests, theologians and philosophers of the western world, as the arbiters of the highest and most important truths.

It is not surprising, then, that, since that (seeming) triumph, they have had *little wish* to relinquish their glittering prize, by admitting that, despite all the claims for Darwin, we *still* do not have a clue, *scientifically* speaking, as to how Life *really* arises, second by second, *now! and now! and now!,* let alone ***why*** it does so, nor, what drove *those massive changes,* across vast time, which we see in the fossil records.

But, and this is *key,* not only does the (current) scientific community *not wish* to admit to its lack of knowing in this field, *it doesn't need to.*

It is not that, if they are wrong in their claims—"Oh, all you need is some CHEMICALS, some lightning, and voilà!, Life!"—that a plane is going to crash or a new drug is going to kill people. Darwin getting it wrong isn't going to hurt anyone. It's not that kind of science.

But, because it is the *same* people, the very clever scientists and technicians who, now, have so much prestige in other fields—electricity, radio, radar, x-rays, cell phones, TVs, PCs, AI, medicines etc—who have repeatedly assured us that Darwin *really did* solve the mysteries of Life's history, and that the Origins of Life are, they say, *non-supernatural,* processes, that it has become *such an uphill struggle* for those of us who are more skeptical to question their *'wisdom.'*

They have been right about so much else, so why not about this? Yet, again, we forget that our current origins science *does not depend* for its success *on being right* on important TECHNICAL things, like whether a Mr Darwin's plane will fly, or his drugs will be safe, but on ***its ability to persuade us*** that what it tells us about origins is true.

No one will ***persuade us to get on a fatally flawed plane*** or to take a *dangerously unproven drug*. But if we step into Darwin's time machine, the worst, seemingly, that can happen to us is that we may end up believing in a whole lot of nonsense: "Oh, we descended, by pure *stupid* chance, from NON LIVING CHEMICALS, which, accidentally became bacteria, then eventually us, for no reason at all." If Darwin's *Just So* story appeals to us, we can believe it. If we cannot accept that the old religious creation stories were never meant to be *literal* descriptions of *creation,* but, rather, to symbolize to us that our *Lives and our Loves,* despite their many challenges, have *intelligent sources,* not mindless, then we may go for Darwin's, "It's all just a *chemical accident,"* model.

It is not a Life or death choice. What so grates, though, is the claim that it is *'scientific'* to believe that, 'Life is just a series of CHEMICAL ACCIDENTS, there is nothing *supernatural* to it,' when this so clearly doesn't work, any more than it is right to claim that religious creation stories are *'scientific.'*

The huge difference being that the old religious creation stories were only ever intended to be *metaphors,* whereas Darwin's *Just So Story* was meant to be more than just a creation story for the atheistic. "No," its fans tell us, "it is *literally* true! We can prove it to you, Life is just an accident, which evolved by accident! This is just a fact, like gravity!"

None of us should have a problem with scientists, like Darwin, trying to make sense of Life's origins. What, after all, is more important than knowing ***Who*** and ***What*** we really are? Are we just cosmic accidents or are our existences more *meaningful* than that?

The problem is not our desire to explore, but our dogmatic attempt to explain all of *Life, from Cells to Shells, Worms to Fish, Birds to Bees,* you and me, even *Consciousness and Mind,* as if all these amazing things can be reduced to mere FLUKES OF CHEMISTRY.

This is not the zenith of science but its nadir. It makes science look foolish. It was not wrong, for Darwin, to *try* to explain all things in PURELY PHYSICAL terms. It was a daring idea. What if it was true?

What if we really are alone in the universe? At *all levels* and *dimensions?* What if there really is no *God or Spirit?* No *Soul* or deeper meanings to anything. But, once it became *clear* that Darwin's approach didn't work, we should have moved on—not clung on to his ideas, now so long past their sell by date. To say, "Oh, the scientific enterprise has to treat every question as if it *must* have a purely *physical* explanation," as methodological materialism and methodological naturalism (MM/MN) require scientists to do is question begging.

What if it's not true? It is only if we assume that the only possible *'scientific'* explanation is a MINDLESS-CHANCE one, a purely PHYSICAL one, that we will, eventually, paint ourselves into a mental corner, as thinkers like Darwin, Dawkins, and Lewontin have all done.

They are so sure that the *supernatural* cannot exist, that it makes it *impossible* for them to concede that there *may* be other, better explanations than theirs. That there may, after all, be far more to reality than they dream of in their *limited, science-confining* philosophies.

But, because thinkers, like Darwin, who are interested in Life's origins, are under *no practical pressure* to make any such concessions, they can, pretty much, say what they like, whether it is realistic or not.

This is why it is so important for us to understand that, despite all the claims for Darwin and his theory of Evolution, over nearly seventeen decades now, this topic is much less about good, proven science, than it is about the battle in our culture as to who is best qualified to explain reality—overall—to us. Is it our philosophers or our physical scientists? Our meta-physicians or our tech-nicians?

What have our scientists, engineers and technicians given us? So many *practical* things, too many to count. What do our philosophers, priests and meta-physicians give us? At best, good moral frameworks, belief in *Spirit and Soul,* deeper meanings, a love for *Spirit* and *Creation,* and for our Fellow Beings, hope in a Life or Lives to come.

On origins, what do the two schools give us, today?

One school—the clever, *technical* school, so good at manipulating matter, and the forces which govern it—now mostly tells us that only the PHYSICAL WORLDS of atoms and molecules are real and that C*onsciousness, Mind, Soul, Spirit,* do not exist in their own right. They may *seem* real but, really, they are but the random byproducts of accidental combinations of mindless *atoms* which eventually formed the first bacteria, then the first worms, spiders and fish, and, eventually, you and me. 90% + of the scientists in the American National Academy of Sciences, the NAS, belong to this modern, materialist-atheist [96] school.

They do not agree that Life may have *supernatural or spiritual* origins. No, they go for Darwin's, "Mindless Particles to Clever People by Pure Chance," theory. Yet, does anyone's expertise in PHYSICAL science, amazing as it is, give them the authority to claim, in the name of holy reason, and sacred logic, that they are *scientifically* justified in insisting that only the PHYSICAL REALMS can be real, and, therefore,

"Life ***must*** just be an accident which evolved by accident?"

The main reasons for thinking in this way are, as far as I can tell, (a) that Life's *deeper, intelligent, sources,* if real, are not physically visible, (b) that the old religious creation stories are not *literally* true, (c) that Darwin had an interesting theory, and (d) that, "If we, now, admit that his ideas were not, in the end, borne out by the real-world evidence of the fossils etc, we will look foolish, and we may lose some of the huge status accrued to us, based on our, previously, far too confident claims that we had successfully solved some of Life's deepest riddles."

How, "Dead chemicals," we said, with far too much confidence, "had become stunningly complex DNA CODES for no reason, Genes for no reason, Cells for no reason, Hearts for no reason, Blood for no reason, Brains for no reason, Worms and Fish, Spiders and Snails, Bears and Whales, for no reason, Apes and People, for no reason…"

This dismal, tottering, rotting structure of Darwinian speculation upon Darwinian speculation, which would not have lasted for five minutes, in other more ***practical*** fields of science, involving real matters of Life and death, like safe airplanes or safe drugs, will come crashing down. This is why there is a lot of pressure to continue to pretend that our physical-science friends have successfully explained even the very origins of Life, when they have done no such thing.

We all know that there are many different schools of religion and spirituality—*Hindu, Yogic, Shamanistic, Buddhist, Jewish, Christian, Islamic, Theosophic, Cabalistic, Anthroposophical, Spiritist, Spiritualist* —and that, on many things, they disagree, but, on this one thing, they do, *all* agree: Life on Earth is not just some lucky cosmic accident, but it is caused, and it is driven by, *vastly intelligent Life, Soul, and Spiritual Forces. Forces which are as real as,* or perhaps even more real than, anything else, even if they are not as obvious as those driving outer PHYSICS and outer CHEMISTRY. But, without those *deeper, more subtle forces,* nothing would work, let alone work *intelligently.*

It is worth briefly visiting, again, a simple but useful way of looking at the world which our science has abandoned, to its intellectual cost, and to our *spiritual* cost: philosopher Aristotle's famous four causes.

He said that everything in Nature has four causes: (1) Final Cause—its *teleology* or its *purpose,* what it's for. (2) Formal Cause—its *shaping idea,* its form. (3) Material Cause—what it's *made of.* And, lastly, (4) its Efficient or Mechanical Cause—how it's put together and *how it works.*

Take a Car. Purpose: transport. Shaping ideas: body, wheels, motor, seats etc. Material cause(s): steel, copper, rubber, plastic, glass. Mechanical causes, motor, gears, wheels. Aristotle's idea, although simple, does work as quite a useful way to make sense of things.

However, in the 1600s, at the dawn of the modern scientific age, some thinkers, like Sir Francis Bacon, began to argue that while we can work out what many things in Nature are made of, and how they work, we cannot, nearly so easily, say what they are *'for.'* Fair enough.

For example, if we are thinking in strictly *physical-scientific,* rather than in *spiritual* terms, how can we say, with any confidence, what a worm, or dog, or horse, or human is 'for?' What it's 'made of' (materials) and 'how it works' (mechanisms) which we are now so good at working out, don't tell us the *'why'* or the *'how'* of how something comes to be in the first place. To which second questions there are only two possible answers: either it arose by pure, DUMB CHANCE or it arose due to an *intelligent cause* or causes of some kind or kinds.

Spiritually, we may have some ideas or answers to these profound questions, but how can we know from a purely PHYSICAL SCIENTIFIC analysis of materials and mechanisms? This led thinkers like Bacon to argue that it was futile for the natural philosophers, or scientists, as we now call them, to try to work out Aristotle's *formal* and *final* causes, the ***how,*** and the ultimate ***why,*** of things, interesting as it was, but to leave that to the *philosophers* and the *meta-physicians,* to the *priests* and the *theologians,* and to focus, instead, more seriously than ever before, on Aristotle's last two causes, *"**made of**"* and *"**how it works.**"*

Bacon described his proposed process in a rather unpleasant way. He said, we must "put Nature to the rack," to torture her secrets from her.

This, then, is what his followers proceeded to do, and it is thanks to their efforts to unravel Nature's secrets in this highly focused, if potentially brutal way, vivisection being one of its horrendous by-alleys, that we now have so many useful drugs, gadgets and technologies today.

The problem with looking at Nature in this ruthless way is not that it isn't powerful, but that it only looks at the purely MATERIAL and the purely MECHANICAL components of the Natural worlds—at what things are *made of* and *how they work,* while ignoring the ***how*** and the ***why*** of Nature's *deeper, Soul and Spirit* components. It is all about the beautiful BODY of Nature, nothing about her *Soul,* or her *intelligent causes.* And it certainly doesn't tell us how *Living things* arise in the first place.

Desert islanders could, perhaps, work out, to a degree, what a never-before-seen Car, is *made of* and *how it works.* Similarly for Dogs and Horses. We can work out what they are *made of* and *how they work,* and we can use, or abuse, them in various ways. But, just because we can do this, it doesn't mean they *have no deeper* causes.

However, with so many successes, Bacon's natural philosophers, our scientists and technicians of today, began to forget that his point was *not* that there ***are*** no formal and final or *intelligent '**why**'* causes in Nature, just that we don't need to ***know*** what they are, in order to do *useful, practical science* on them—around *'**made of**' and '**how they work**'*— leading to so many *useful and practical results,* like Cars, TVs and PCs.

Abandoning Aristotle's first two causes, the *purposes,* and the *ideas* behind things, the *intelligent* and *meaningful **Why**?* of things, is not an issue for ***practical,*** everyday science. But, it becomes a huge issue when we are looking at Origins. Because we then have to argue, as thinkers like Lewontin do argue, that nothing in PHYSICAL Nature is, or can be, caused by anything *beyond* the laws of PHYSICAL Nature. Yet, at the same time, our materialist friends, like Lewontin, say that the laws of nature are totally MINDLESS and AIMLESS.

This means we have to rely, not on an *inherent, cosmic, intelligence* which ***just is,*** but on PURE, DUMB, LUCK, which ***just is,*** to try to make sense of Nature's amazing creations, including of we, who are able to think and to *ask intelligent questions* about it all.

Our desert islanders don't need to know *Who?* made the Car, nor *Why?,* to work out how to use it. Similarly for us when it comes to dogs or horses. But, this approach only provides us with ***half*** an explanation, which, for everyday purposes, is fine. We don't ***need*** to know *Who or What* made the Cell, Car, TV, Dog or Horse, but *as thinking Beings,* with a love for Truth, we love to know *what is True*—for its own sake.

So, could this or that amazing *Living Being* be caused by blind chance, without any *spiritual (or why?)* causes, as Darwin argued`? Or, must it, also, have formal *(intelligent)* and final *(purpose)* causes as Aristotle's, and pretty much everyone else's, thinking required, before Darwin's "Life is probably just a chemical accident" ideas took off.

While some of us are more mechanically minded and more interested in how things *work* and *how to use them* than in ***who*** or ***what*** made them, or *why,* others of us just want to know *what is true,* even if it makes no *practical* difference. On our desert island the more practical have worked out what the unknown Car is *made of, how it works,* and *how to use* it. But they are not so interested in *who* made it or *why.*

The island's *philosophers,* less concerned with its *practical* aspects, agree that the mysterious machine can only have one of two possible causes: either (a) it's just a fascinating accident of its various parts, which came together by pure chance, or, (b) there is *intelligence* behind it, which, if true, is fascinating. Because, *where* where is the intelligence that made it? "This fascinating machine certainly ***looks*** as if it is designed, not just some random thing. But *where are its designers?"*

The more *practically minded* have, meantime, taken the car apart and worked out lots about it. They tend to take the view that, given that there are no intelligent designers *anywhere to be seen,* the Car probably just arose by natural chance—the laws of probability + time created it.

Perhaps the cars parts *'self-organized'* due to nature's regular laws. But regular laws, the philosophers point out, can only create repeating patterns as e.g. we see in crystals. They cannot write the totally **specific**, kind of software CODE required to make a complex machine run.

The practical ones disagree, "No, it's probably just another 'natural' phenomenon, like the wind and the rain, which naturally evolved due to the mindless laws of nature plus the mindless laws of chance." The *philosophers* continue to question. They say, "If it ***looks*** intelligently designed, then despite the fact that its intelligent designing forces are nowhere BODILY to be seen, then it probably ***is*** intelligently designed."

Yes, it's just a metaphor, but it sums up the state of our origins debates today. Did Life arise, and evolve from there, due to nothing more than the regular laws of physical nature + lucky chance? Or is an *underlying intelligence* involved, in the 24/7 functioning of our *trillions* of cells, in our ever circulating blood, and ceaselessly beating hearts? Which of these two most basic choices makes the most sense?

9 Fabulous Chance or Fat Chance?

In *The Blind Watchmaker,* Richard Dawkins stated that, "Biology is the study of complicated things that give the *appearance* of having been designed for a purpose," but, he said, "the evidence of [Darwinian] evolution reveals a universe *without* design."

For Dawkins, like his hero, Charles Darwin, all things, even our *very minds,* can be explained by totally MINDLESS NATURAL processes. This seems illogical to me, but it is why Dawkins writes, "Darwin made it possible to be an intellectually fulfilled atheist." [97] Darwin, he feels, showed us that materialism is *scientifically* justified.

"Move on *spiritual* people, you're kidding yourselves."

Yet, can our blood and our faithfully beating hearts, can our very *minds,* can our capacities for *knowing, thinking, and creating, really* be reduced to *mere coincidences of chemistry?* Is this way of thinking *honestly* scientific?, let alone intellectually fulfilling?

In his foreword to the 20th Anniversary edition of Philip Johnson's bestseller, *Darwin On Trial,* Michael Behe notes that,

> "the situation for materialistic [atheistic] origin-of-life theories has gotten substantially worse [since Johnson first published]. Broadly speaking, for decades there have been two categories of [materialist no-intelligence-was-needed-it-arose-by-chance] origin-of-life theories: [a] the "metabolism-first" view, where metabolic reactions in an enclosed space precede the occurrence of genetic material; and [b] the "genetics-first" view, where a DNA-like polymer that is capable of carrying information precedes cells. The partisans of both camps have offered *devastating criticisms* of each others views, so that none are left standing... the *only reason* at present to believe in a MATERIALISTIC [or mindless] origin of life is if one holds it

as a postulate that life *must* have had a [purely] materialistic [non-spiritual] origin." [98]

However, the Darwinian-type idea of "a purely materialistic origin" of life, in a 'little pond,' doesn't work, because, as Dembski explains in ***No Free Lunch: Why Specified Complexity Cannot Be Purchased Without Intelligence,*** there is no such thing as *'free'* software. Yet: all living things are run by this same, mysterious, DNA CODED software, which makes our most sophisticated IT look *basic!* The counter claim to this is that our star, the Sun's energy, which is *'free,'* could overcome the disorganizing forces of *entropy* and, by pure *chemical luck!,* create life's CODEs, *Cells, Shells, Teeth, Blood, Hearts, Kidneys etc.*

I hope, though, that you will immediately agree that this proposal is even worse than the idea that some monkeys could randomly type some *'free'* Shakespeare, that a wild fire, with lots of *'free'* fuel, could generate anything other than burnt wood, or that a tornado blowing through a junk yard could assemble a modern, F35, fighter jet.

One of the ironies of this debate, dear reader, is that those who argue for the MATERIALIST view, in the name of rational, logical, science, never stop to consider just how many *nonsensical* things they must believe, *"all before breakfast,"* as Lewis Carroll might have put it, just to avoid the ancient insight, that *Being and Beings—Cells to People, Atoms to Stars*—arise from astonishingly *Intelligent Forces,* which ***'just Are,'*** rather than from totally mindless forces which must also ***'just be.'***

Which makes more sense, DUMB universe or *intelligent* universe? Which, of the two, is the more intellectually satisfying? Interestingly in a letter to the astronomer John Herschel, Darwin wrote,

> "One cannot look at this Universe with all [its] living productions & man without believing that *all has been intelligently designed;* yet when I look to each individual organism, I can see no evidence of this." [99] [Ah, dear Darwin, you cannot have been looking very closely! Try making just one cell, or heart, or eye!]

Sir Frederick Hoyle said, *Nature's* mechanisms imply "higher intelligences" and that this should be "self-evident." Ah *yes,* and how true, and if this area of science were something *practical,* like whether Darwin's will actually fly, the debates would be much shorter!

No one would want to fly in Charles Darwin's dodgy plane. But because origins science is not about ***practical*** knowledge, of what *really works,* it can make up *'just so'* stories, all day long, about how Life ***'might'*** form by totally 'mindless chance' processes and say,

"Oh, didn't you know, scientists have now proved that Life could start by pure LUCK, because they know a lot about the solar system, that Life is very old, and *no intelligent* input was needed!"

'Hold on, that's not true. At present we have no good *scientific* ideas as to *why* or *how* life arises.' But, taking no notice, they reply, "Life is just a *chemical fluke,* and anyone who disagrees is ignorant."

No, they are conflating *science,* which is meant to be an open exploration of *all* of reality—be it solely PHYSICAL or be it also *Spiritual*—with their *belief system* which says that unless it's PHYSICAL, and *physically* measurable, it cannot exist. Not noticing that we cannot *physically* see our ideas, our feelings, or our minds.

Not noticing that we cannot *physically* see love or intelligence, compassion or strength, yet all are utterly real. This dreary ideology which dismisses the *spiritual* out of hand, even as it insists that such stunningly complex things as *Cells!, Hearts, Eyes!, Brains!, Minds!,* could arise by pure *luck!,* when nothing could be less true.

Yet, thanks to Darwin, when we observe a soaring Bird, a swooping Bat, or a graceful trapeze Artist, we are to agree that these are but the *purposeless* and *meaningless* products of a wholly mindless process—and we are to persuade ourselves that this is *"intellectually fulfilling!"*

Any academics who dare to question Darwin's unrealistic *(impossible is better) paradigm of unintelligent design* put their careers at risk. I find it odd, though, and I trust that you do too, that scientists today are thought to be *clever enough* to work out that nature is mindless to its core, without *meaning* or *Soul,* as Darwin argued, but *not* clever enough to work out that this may not be true. The very opposite in fact. That *Nature* is *pervaded by intelligence and design.*

It is also sad that most of those who follow these arid *Soul and Spirit denying* ways of looking at reality, refuse to take any interest in the extensive modern evidence[100] for *mind-beyond-the-body* phenomena, like near death, out of body, and other spiritual experiences.[101]

Evidences which amply falsify materialism. Their minds are closed, they don't to want to know—that their Darwinian Emperor has no clothes. But it is one reason why our current science culture lacks the intellectual curiosity to understand that an all-pervading, self-existing, *Spiritual Intelligence,* which ***just Is,*** might express creation in many different ways, not just through our *Souls,* but through the *Souls* and *senses* of many other organisms, all with their own unique qualities of *Sentiency* and *Awareness.*

Did you know, dear reader, that there are dogs who enjoy skate-boarding and parrots who literally like to dance, in good, rhythmic, time, to our music. You can see delightful clips of these online. Who are we to say that these fascinating *Beings* have no *Soul* or reasons for existing, just because we don't know exactly what they're *'for'?'*

Michael Denton says that, for us to build a scale model of *just one, tiny cell,* at one, tennis-ball-sized atom *per second* would take us *hundreds of thousands of years!* Yet, each *Cell,* of which there are *trillions* in our bodies, achieves this at supercomputer speeds, *in just a few hours*. Darwin didn't know these things, but *we* do.

John Lennox, emeritus professor of mathematics at Oxford, writes that "(1) Life involves a complex DNA database of digital information. (2) The only source we know of such language-like complexity is *intelligence*. (3) Theoretical computer science indicates that unguided chance and necessity [natural laws] are incapable of producing [this kind of] complexity." Intelligent design theorist, Werner Gitt agrees: "It has never been shown that a CODING SYSTEM and semantic (meaningful) information could originate by itself… The information theorems predict that this will never be possible. A purely material origin of life is [thus ruled out]." [102]

Lennox adds, "One would have thought that scientists would prefer an explanation [like intelligent causation,] that explains a given phenomenon over an explanation [mere luck,] that does not.

"The fact that this is not the case, when scientists are thinking about the very origins of life, shows us that ***an a priori materialism*** can produce ***a profoundly anti-scientific attitude*** – [and an] unwillingness to follow evidence where it clearly leads simply because one does not like the [*spiritualist*] implications of so doing." [103] [Sorry Lewontin].

Oh yes, especially as the debate here, is *not* about practical science, which doesn't need us to think about *Life's deeper* causes.

We know how to make electricity and how to use it. Whether Life's origins are *intelligent* or RANDOM has no impact on this. So we can call this kind of science, ***practical*** or ***technical*** science.

But, what we can refer to as ***philosophical science,*** is much looser because it doesn't *need* to be so rigorous. Our discoveries in natural history, like the Cambrian Big Bang, the Dinosaurs, or our ideas on whether Darwin was right do not, *practically,* have to be *'right.'* They are not matters of Life or Death, like whether a plane will fly or crash.

Yet, while not *practically* important, the *spiritual* stakes are, ironically, much higher. Whether Life has *meaningful* causes, as most before Darwin held, or RANDOM, has *huge* implications, for our relationships to each other, and for our relationships to Existence itself.

Are our *spiritual insights* real or not? Are there *Higher Power(s),* we can *inwardly* turn to, or not? Are our *mystical* experiences, including *knowings of Oneness, of heart-felt merging with the Whole, of realizing the Essential Goodness of Existence, reflective of a deeper, Spiritual reality, or not?* If materialism is true, the answer to all these questions has to be "no." If materialism is false, the answer may be "yes."

But given that so many of us now see *'science'* in opposition to all things *spiritual,* how will this war between religion and science ever end? Well, the real fight is not between religion and science, but between the *materialist* and the *spiritualist* worldviews. The materialist camp tries to persuade us that *it* owns 'SCIENCE.' It doesn't.

Science is but a series of methods for investigating the *whole* of reality, the *all of 'what is,'* be it purely **PHYSICAL** or be it, also, *spiritual.*

As far as the religions are concerned, while they can, all of them, be unreasonable at times, their *basic claim* that Life's ultimate causes are *intelligent,* not **DUMB**, is logical and reasonable.

A similar realization, arrived at by modern, *non-religious* means, confirms their ancient insight. But that is as far as it goes.

Modern observations of *intelligence and design* in Nature confirm *some* of our ancient intuitions about Life's deeper sources, but we *cannot,* nor do we need to, go from reasoned, logical admissions of *intelligence* and *design* in nature, to endorsing a particular religion, its beliefs about heaven and hell, its recommendations for salvation and so on.

10 Genetic Copying Errors

If we remain open to the classical idea that *Nature* encompasses *subtle, non-physical, spiritual aspects,* we can allow for intelligent causation. But if we rule this out, from the first, we will be forced to *try* to explain all elements of *Life,* even our very *Minds,* as though they are purely PHYSICAL things, arisen by pure chance, as materialism asks us to do.

"But surely," you may say, "we know, very well by now, unless, for some emotional or political reason, we would rather not know, that blind luck could *never* create a *microbe* in the first place, let alone evolve it into a *Worm, Snail or Spider,* and, later, *amazing Beings* with the minds and instincts of *Eagles, Bears, or Humans.*" Yes, we can agree on that.

We also know very well, by now, that a species can only vary *within,* but *not beyond,* the limits set by its gene pool. There is *no data* for the extreme levels of species *changeability* Darwin called for—*microbes to Men!, particles to People!,* let alone due to unintelligent *chance.*

"What about *wolves to dogs?* Isn't that an example of Darwinian Evolution?" No, because no dog, not even the tiny *Chihuahua,* is truly *a new species.* Such *intra-species* variations do ***not*** explain (a) how Darwin's theoretical common ancestor microbe might arise in the *first place,* nor, (b) how, later, it might turn into VASTLY LARGER, and vastly DIFFERENT Snails, Bears, Whales, Apes, or People.

Note, too, that the modern version of Darwin's theory—neo-Darwinism—which attributes such shockingly unrealistic powers to *random genetic typo-mutations,* does *not* explain where the original, non-mutated genes came from in the first place. It just blusters that *'Oh, you know, they must have evolved, somehow or other,'* and it hopes that we are not paying attention.

Darwin didn't know about genes. But, some time after he published, Gregor Mendel's earlier work on genes was re-discovered. It showed how genetic variations *within* species arose, and how this shuffling of *existing* genes, within *existing* species, had been *intelligently selected,* by clever human breeders, in their pursuit of improved varieties of dog, pigeon, or tulip.

In the wild, it was this natural shuffling of *existing* genes in *existing* species, which favored the *superior survival rate,* the 'natural selection,' of those species members with the gene mix best suited to the prevailing conditions, for example, finches with bigger beaks, or the fittest bears.

Mendel's discovery did *not explain*, though, how a microbe, worm, or fish, might (slowly) turn into an *entirely other* kind of life form, as Darwin predicted: microbe-to-worm (vast change), worm-to-fish (vast change), fish-to-dinosaur (vast change), dinosaur-to-mammal (vast change), bear-to-whale (vast change), ape-to-human (vast change).

Mendel had simply worked out how a species *existing* gene pool allowed for minor, *but strictly limited,* variations *within* that species.[104] He had not discovered anything to say that an existing species, let alone an entire *phylum,* could be changed—*by chance* or *by people*—into a completely new one, via, for example, a sea-going stage, like fish, then land-going, like dinosaurs, then mammals *(all chordates),* then mammal carnivores *(order of chordates),* to a carnivore family *(canidae),* to genus *(canis),* to wolves, *(canis lupus),* as Darwin predicted.

However, after the discovery of Mendel's earlier work on genetics, some people wondered whether ***random genetic mutations*** (RGMs), which are rare ***genetic copying mistakes,*** the effects of which Mendel did not explore, might not generate *new* genetic information, and lead, eventually, to *new* plants and animals, which—*if this was true*—might make Darwin's ideas more plausible.

This idea is called neo-Darwinism because Darwin did not know about genes. RGMs are, however, *copying mistakes* in the genetic code, which, we now realize, *Cells, very actively, protect against.*[105] However, when RGMs do, occasionally, escape the *Cell's* very clever, self-correcting mechanisms, they are *generally harmful,* sometimes lethal, and no more do they provide useful new constructional information than a poorly copied microchip would be a useful new feature in a computer.

In ***God's Undertaker: Has Science Buried God?*** Professor John Lennox tells us how the "incredibly precise duplication of DNA is not accomplished by the DNA alone: it depends on the presence of the living cell." Lennox quotes biologist James Shapiro who writes,

> "It has been *a surprise to learn* [just] how thoroughly cells protect themselves against precisely the kinds of *accidental genetic change* that, according to conventional [Darwinian] theory, are the [very] sources of evolutionary variability. By virtue of their *proofreading* and repair systems, living cells are not the passive victims of the random forces of chemistry and physics. They devote large resources to *suppressing* random genetic variation and have the capacity to set the level of background localized mutability by adjusting the activity of their repair systems."[106]

"So it is unlikely that *random genetic typos* could drive the incredibly progressive evolutionary processes Darwin had in mind?"

Actually, it is *beyond* unlikely. Yet, currently, any scientists who point out that these very clever error correction processes point, not to blind *'luck,'* but, to an *amazing level of intelligence* in living things, risk being told that they are being fanciful and *unscientific!*

Don't you find it odd, though, that in our supposedly so very rational, logical, culture, something as obvious as this is ignored? Despite the fact that these *error correction processes* in cells ***falsify Darwin's entire theory***—of descent with vast modifications: bacterium to worm (VAST change), bacterium to spider (VAST change), bacterium to fish (VAST change), bacterium to snail (VAST change), alga to oak (VAST change), bear to whale (VAST change), caterpillar to butterfly (VAST change), land-running rodent to fast flying bat (VAST change).

The cellular *error correction processes* PREVENT from happening the very thing which Darwin's modern followers say are the very engine of the evolutionary process he envisaged. No, *random genetic mutations* marketed to us as the tools of an astonishingly progressive evolution—

Particles to People—*cannot build the new,* and any minor advantages, which they occasionally confer, always come at a cost.

In ***The Edge of Evolution: The Search for the Limits of Darwinism,***[107] biochemist Michael Behe explains that *random genetic mutations are not constructive.* For example, the unpleasant *sickle cell* mutation is *not* desirable, even if it does have the one benefit that it can confer some immunity to malaria.

Generally, though, random genetic mutations are *harmful,* and lead to the *loss* of genetic information, *not to construction,* let alone to new species, never mind new *phyla (new body plans).* In fact, although Darwinism can explain *minor* differences in organisms, like different varieties of pigeon or dog, RGMs could not possibly drive the very powerful transformational processes Darwin had in mind.

Michael Behe adds that there is *no data* to support Darwin's idea that the barriers, between the species, up to the level of families, could ever be breached by any kind of random process.

In fact, despite their intelligent, focused efforts, clever human selectors have never been able to cross even the species barrier. Chihuahuas remain, in essence, *wolves,* bacteria are still *bacteria,* tulips are still *tulips.* Pigeons stay *pigeons,* and Fruit Flies remain *fruit flies.*

Dr. Durston, a scientist writing in *Evolution News* notes,

> 'There is mounting evidence that most, if not all the key predictions of the neo-Darwinian theory of macro-evolution are being *falsified* by advances in science. ... Ask computer programmers what effect ongoing *random changes* in the Code would have on the integrity of a program, and they will universally agree that it *degrades the software.* ... deleterious mutations [do not create biological life, they destroy it].' [108]

Despite all the pro-Darwin hype, for more than 160 years now, random genetic mutations are *not beneficial,* and, far from causing a genetically deformed creature to be turned into a new life form, would lead to it being *eaten!* Even antibiotic resistance in bacteria is caused by genetic copying mistakes, which are corrected out of the population when the antibiotic is no longer used. In *Darwin Devolves,* Behe notes that Darwin did *not show* that living mechanisms, be they microbes, or organs like livers, eyes, or brains, *really could* be built by natural selection. He just argued that it *might* be possible.

Behe also points out that, while influential thinkers like Sir Francis Bacon suggested, at the dawn of the modern scientific era, that scientists need not worry about the Aristotelian *'purposes'* of things, be they Atoms, Rocks, or Horses, in order to do good, practical, science on them, to deny that anything in nature has *teleology,* or purpose—as scientists are *currently* trained to do—is so clearly, untrue. As Behe says, while we may not know what a Horse, Dog or Human is ultimately *'for,'* we *do* know that its heart and its lungs, its mouth and its limbs, all have *teleologies* or purposefulnesses which make the amazing *Living Mysteries* which we call Horse, Dog, or Human, possible.

Behe notes that RGMs, which are held to be the very driver of Darwinian Evolution, will only be naturally selected, and spread through a population, if they confer a benefit. But, he says, this *always* comes at the cost of the *loss* of some other genetic information.[109]

RGMs do not build, they occasionally bestow minor benefits, but, generally, they *degrade.*

This is why the idea that a dinosaur foreleg could gradually turn into a bird's *wing, with its amazing feathers,* due to a random series of *genetic typos*—which all cells actively *protect against*—is no longer good science. Actually, *it was never good science.*

James Perloff, author of ***Tornado in a Junkyard,*** a title referring to Frederick Hoyle's comment that it is as likely that Life could arise by chance as it is that a hurricane blowing through a junk yard could assemble a *Jumbo Jet,* writes, "According to Darwinism, single cells eventually evolved into invertebrates.., then successively into fish, amphibians, reptiles, and finally mammals. Darwin said this occurred from creatures adapting to environments. . . . [But Mendel's] discovery of genetics threatened this claim. New organs require new genes."

However, as Perloff says, merely "moving into new environments doesn't give you new genes." This initially stumped Darwin's fans.

Then, they thought they had found a solution. "Random mutations—*copying mistakes* in the genetic code—occur very rarely, but [they] DO alter genetic information." So, although Darwin didn't know about genes, his later followers said, "animals gained new genes by chance mutations…which they adapted to evolve into higher forms."[110]

However, Perloff comments, *"chance mutations* are to the genetic code what *typos* are to a book: they *remove* information, but *do not improve it.* In humans, mutations cause sickle cell anemia, cystic fibrosis, hemophilia, Down's syndrome, and *thousands* of other genetic diseases . . . even the rare "beneficial mutations" evolutionists trumpet – such as bacterial resistance to antibiotics – actually result from functional [genetic] losses."

Let's take a look at the *feathered* wings of Birds and the *featherless* wings of Bats. Both appear in the fossil records *suddenly,* and *fully formed,* in the usual, *non-Darwinian—non-evolutionary—way.*

There is ***no fossil data*** to support Darwin's idea that Birds or Bats wings evolved from wingless dinosaurs for birds, or wingless mammals for bats. As Michael Denton explains in *Evolution: A Theory in Crisis:*

> 'It is not easy to see how an impervious reptile's scale could [as Darwin's supporters claim] be converted gradually into an impervious feather *without passing through a frayed scale intermediate* which would be weak, easily deformed and still quite permeable to air. It is true that basically the feather is indeed a frayed scale – a mass of keratin filaments – but the filaments are not a random tangle but are *ordered in an amazingly complex way* to achieve the tightly intertwined structure of the feather. Take away the exquisite co-adaptation of the components, take away the co-adaptation of the hooks and barbules, take away the precisely parallel arrangements of the barbs on the shaft and all that is left is a soft pliable structure ***utterly unsuitable*** to form the basis of the stiff

> impervious feather. The stiff impervious property of the feather which makes it so beautiful an adaptation for flight, depends basically on such a highly involved and unique system of co-adapted components that it seems ***impossible*** that any transitional feather-like structure could possess even to a slight degree the crucial properties. In the words of Barbara Stahl, in Vertebrate History: Problems in Evolution, as far as feathers are concerned, "how they arose initially, presumably from reptiles scales, ***defies analysis.*** " [111]

Darwin's supporters try to counter this by saying that birds feathers must have arisen—[due to genetic copying mistakes]—from reptile scales because they are made of *similar* materials. However, would we argue that airplanes *must have* evolved from cars just because they are made of similar materials? And by pure chance?

Darwin's fans also argue, based on his theory, and the fact that there are some similarities between dinosaurs and birds, that birds lungs evolved from the standard vertebrate lung. But, as Michael Denton says,

> 'In all other vertebrates [including dinosaurs] the air is drawn into the lungs through a system of branching tubes which finally terminates in tiny air sacs, or alveoli, so that during respiration the air is moved in and out *through the same passage*. In the case of birds, however, the major bronchi break down into tiny tubes which permeate the lung tissue. These so-called para-bronchi eventually join up together again, forming a true circulatory system so that air flows in one direction through the lungs. This unidirectional flow is maintained during both inspiration and expiration by a complex system of interconnected air sacs in the bird's body which expand and contract in such a way so as to ensure a continuous delivery of air through the para bronchi. ... NO LUNG in any other vertebrate species is known WHICH IN ANY WAY approaches the avian system.

> 'Moreover, it is identical in all essential details in all birds as diverse as hummingbirds, ostriches and hawks. *Just how* such an UTTERLY DIFFERENT respiratory system could have *evolved gradually [due to countless genetic typos, which all cells protect against!]* from the standard vertebrate design is fantastically difficult to envisage, especially bearing in mind that the maintenance of respiratory function is absolutely vital to the life of an organism to the extent that the slightest malfunction leads to DEATH WITHIN MINUTES.' [112]

Yes, dear reader, *death within minutes.* We need to think. Perloff points out similarly disastrous flaws in Darwinian thinking when it comes to our understanding of life at the *cellular* level. He writes that, for a Darwinian or materialist account of Life to be true,

> "the primordial cell must have perfected—in the span of one lifetime—the process of cellular reproduction; otherwise there never would have been a second cell." Yet, he says, despite the demonstrated I M P O S S I B I L I T Y of such a process, people are still being "taught that life began [*not* due to an intrinsic intelligence, which *just Is,* but] from a fortuitous [yet scientifically impossible] arrangement of chemicals [which just happened!]."[113]

Perloff adds, "According to Darwinism, single cells slowly evolved into multi-celled invertebrates (creatures *without* backbones, like worms, snails, jellyfish), then successively into [creatures *with* backbones, like] fish, amphibians, reptiles, and finally mammals. Darwin said this occurred from creatures adapting to environments."

But, he asks, if *"bacteria* evolved successively into *invertebrates,"* then fish, amphibians, reptiles, and mammals, there ought to have been countless "transitional stages." Isn't it odd, he asks, that no such transitional fossils are *ever to be found,* let alone vast numbers of them? "For a Fish to become a land creature, turning its fins into legs would require *new* bones, *new* muscles, *new* nerves—and while it was adapting to life on land, a *new breathing system!* Since this supposedly happened from chance mutations, which are *rare events,* innumerable creatures would have to live and die during the intermediate period." But: "WHERE IS THE EVIDENCE for these TRANSITIONALS?"

> "Not in the living world. Among bacteria, invertebrates, fish, amphibians, reptiles and mammals, there are many thousands of species, but NO INTERMEDIATE SPECIES between these groups. ... Evolutionists try to explain the missing intermediates by saying "they all became extinct." But, NO TRANSITIONAL LINKS are ever to be seen: neither living, nor in the fossil records.
>
> "While the fossils do show [minor] variations *within* species, they do *not* evidence change from one animal group to another as Darwin claimed. For example, while billions of invertebrate fossils exist, fossils illustrating their alleged evolution from simple ancestors are missing." [114]

Many liked Darwin's theory because it was clear, simple and, if we didn't think too deeply, it seemed logical. But, living things are *not simple,* nor do simplicity and logic, by themselves, make something true. The *pre-Galileo* theory of geo-centricity was logical, but, it was false. Similarly, Darwin's theory can account for *some* of the data but there is far too much data that it cannot explain.

For example, *sexual reproduction*: Darwin thought that *all* Life flowed from a microbial ancestor like a bacterium which evolved, and branched out, into a vast tree, representing *all* today's species. Those bacteria would, like other microbes, have reproduced ASEXUALLY.

But, as John Morrison points out in *Evolution's Final Days,*

> "Considering the incredibly small probability of having one entity evolve [by chance], what would be the chances of two entities, CO-EVOLVING next to each other, AT THE

SAME TIME, and being perfect for each other? What would be the evolutionary advantage of moving away from asexuality [in the microbial world] and needing another member of your species to mate with? [Can you think of any, dear reader? I can't.]

"The penis and vagina are common in many animals…[but the] issue is the START of it all. HOW are we supposed to believe that some animal in the ocean, which was asexual, had a [random] MISHAP IN ITS DNA to [begin to] form a penis, and another had a mishap to [begin to] form a vagina … then the two started having intercourse... And one developed a vagina with ovaries [due to random, genetic typos] and another a penis and testicles to start a two sex world. …

"There are a lot of "HOW in the HECK!" questions here. You will hear a lot of … [ungrounded speculation] from the evolutionary community but I want you to *think* with your own mind, and try to make sense of this logically... it doesn't make sense!" [115] [He's right, we need to think with our own minds.]

Yet, it can be hard to question the prevailing views. Most of us, even most scientists, are not experts in this area, and if we are told that something, like Darwin's theory, is true, by the 'experts,' who are we to disagree? Yet, on reading Shedinger's *Darwin's Bluff,* I was reminded of the amazing LACK of evidence for Darwin's *extraordinary claims*: for blind luck to create *Cells and Shells!, Eyes and Ears!, Blood and Brains!, Worms and Moths, Bears and Whales, let alone Consciousness, Emotions, Intelligence, Memory, Curiosity, Love, Joy, Creativity.*

Darwin wrote: "If it could be demonstrated that any complex organ existed, which COULD NOT POSSIBLY have been formed by numerous, successive, slight modifications, MY THEORY WOULD ABSOLUTELY BREAK DOWN. But I can find out no such case."

This was *disingenuous* of Darwin, because the onus was on *him,* and it remains on his *fans,* to show us that it ***did*** happen, not for us to prove that it *didn't!* So, we say, "Dear, Darwin's followers, please can you *properly* SHOW US how *Cell, or Eye, or Heart, or Lung, or Kidney, or Liver* arose by pure, lucky, chance?" "You can't…?"

Darwin could "find no such case" because, either, he did not wish to find one, or, because he hardly bothered to look. Because, in reality, there are thousands of such cases: *Cells, Blood, Hearts, Lungs, Kidneys, etc*. Gunter Bechly's review of ***Darwin's Bluff*** is instructive.

Bechly describes himself as a one time "card-carrying Darwinist, serving as a fossil curator in one of Germany's natural history museums." While there, he was tasked with mounting an exhibition expressly intended to show how Darwin's work was more valid than the works of his critics and other naysayers. He writes that to prepare,

> "I decided to give the [Darwin critical] naysayer books a quick read. Books I had been assured were all froth and foolishness. I soon discovered that I had been misled. The arguments in those pages were *neither* shallow *nor* illogical. Instead, I came to see that it was actually DARWINISM THAT RESTED ON A CAREFULLY CONSTRUCTED BLUFF. …
>
> "What emerges from Shedinger's deep dive into Darwin's private writings is a picture of a man wracked by doubts and insecurities about his evolutionary theory, but also a man not above a good *bluff,* one he sold so artfully [to his Victorian public] that he may even have persuaded himself."[116]

How ironic, then, that Darwin's theory, *itself,* turned out to be a "***Just So***" story, little better than a religious creation story, with the crucial difference that, unlike Darwin's claims, which, *absolutely,* were meant to be treated as *scientific,* religious creation stories were *never intended* to be treated as literal descriptions of creation, but, rather as symbols that *Life (really) is a miracle,* not just an ACCIDENT.

Unfortunately, Darwin, like so many other thinkers of recent times ignored two of Aristotle's famous four causes: the shaping ideas behind things (formal causation), and their 'purpose' (final causation). Thinking as a materialist he *had* to ignore them. He *had* to try to explain all the shapes (formal causes) and the purposes (final causes) of *Living things,* and their organs, like the mammalian *Heart or Lung, Tongue or Eye, the Giraffe's Legs or beautiful Butterfly, without* there needing to be any *shaping ideas or purposes* in nature—which, if this approach actually worked, would be fine. Sadly for Darwin and his fans, it does not work, but it is *why* our sciences, currently, do not, and will not, admit that *anything* in Nature has any *intelligence or purpose* behind it.

Because, as soon as we admit to *purpose* in Nature we have to admit to *intelligence* in Nature. However, as soon as we admit to this, we have to admit to an *Intelligent Cause, or Causes, behind* the *intelligence* in Nature, which means that the *Spiritual is, after all, real,* which means that our high priests of the material world, *our physical-scientists,* will no longer have the final say on the nature of reality, but they will have to *allow some space* once more to the truth claims of those who speak for and about *spiritual matters, our spiritual-scientists.*

This is the current battle between our physical-scientist friends, like Darwin, Crick, and Dawkins, who say there is nothing beyond the PHYSICAL to explore and our *spiritual-scientist* friends who say,

"We will never be able to make *full sense* of nature, especially not of *Living Nature,* nor, indeed, of *reality as a whole,* if we continue to insist, as a matter of principle, rather than proper, empirical, scientific data, that there are no *spiritual elements* to reality, that only Aristotle's first two causes, **material** and **mechanical**, what things are ***made of*** and ***how they work,*** are real, that there are no formal (shaping ideas) and final causes (purposes) in or behind Nature, and only the OBVIOUS, OUTER UNIVERSE CAN BE REAL. No *subtle realities, no inner, or Soul, or Spiritual Dimensions to Existence.*

Having said all this, dear reader, please *do* note that I am not claiming that all spiritual or religious claims about the *spiritual* and the *supernatural dimensions* of existence are equally true. Of course not.

No, just that the *basic* religious and *spiritual* claim that there is an amazing spiritual *intelligence, or, many spiritual intelligences, behind all Living (and non-living) things* is true.

11 What Kind of Evolution?

Mindless and *Random* or Meaningful and Intelligent?

Micro-Evolution or Macro-Evolution?

Evolution, as a word, is, now, used so widely and so broadly, that it can mean little more than *change over time,* versus, sudden emergences, like the Big Bang in cosmology, or the Cambrian Explosion in biology.

So, if all Darwin had meant to describe was 'change over time,' to say that Life had *evolved* would hardly be controversial. To muddle things further, this unhelpfully broad word has two almost *opposite* meanings, describing two *very different* kinds of processes, both referred to as *evolution*—one rA*n*D*0*m, the other ***intelligent***.

Landscapes evolve randomly but our technologies *intelligently.* Early Fords to modern Fords. So, what about *Living Things? Cells? Blood, Muscles, Teeth, Bones? Fish, Snails, Bears, Whales? People?* Which makes more sense? Do they arise r*and*o*m*ly or ***intelligently****?*

Our materialist friends answer *"randomly."* Their philosophy gives them no choice. Although they don't refer to their way of thinking as 'unintelligent design,' or UD, given that they reject *intelligent* design, or ID, what other option do they have? They *have* to argue for blind chance as their supreme creative principle, don't they?

This is why Darwin, as a materialist, *had* to argue for an unrealistic *probability-defying, entropy-violating* rise from the non-existent, to the *functional,* from the simple, to the ever-more-complex, from LIFELESS CHEMICALS, in a 'warm little pond,' to *living, motile cells!* From the *dis-organized,* to the incredibly *clever—all for no reason at all.*

Yet, nothing we make, not even a simple mousetrap, as Michael Behe explained in his best-selling book, ***Darwin's Black Box****,* "falls together" by pure luck. Yet *the most complicated things* we know of, *Living things, do?* Have we lost our *minds?*

If some of us don't like the religions, or those running them, trying to control us by terrifying us with hideous hells which only the most sadistic *'spiritual'* forces could conceive, or, (un)inspiring us with boringly pious heavens, it is understandable. Some of the old religious depictions of the *spiritual realms* may well be wide of the mark. But the idea that DEAD—DULL—CHEMICALS in some 'warm little pond,' could, by PURE—DULL—CHANCE, turn into DNA CODED *Living* Cells—then, Microbes-to-Fish, Fish-to-Dinosaurs, Caterpillars-to-Butterflies, Rats-to-Bats, Bears-to-Whales, Apes-to-Humans—is beyond implausible. It's nonsensical. All our experience teaches us that clever mechanisms *always* have *intelligent* origins.

So, we either follow the classical view, or, we argue, as Darwin did, that *none* of Life's forms, *Cells, Blood, Brains, Eyes, Hearts,* are *intelligently* put together. No, they only *'seem'* to be.

Which, if it's true, means that we cannot refer to our *Hearts* as *'clever,'* or our *amazing Eyes, Ears, Hearts, Hands, Feet or Brains,* as designed, because *design* implies *creativity* and *mind.*

Yet our present science does still *insist* that there is no ***intelligence*** *behind or* ***teleology*** *(purpose) to,* living things. Never mind that we, as human *egos,* cannot make our *Hearts* pump, turn our food into *Energy or Muscle, Heal* our cuts and scratches, let alone operate *just one* of the *trillions* of stunningly complex *cells* inside of us.

Our technologies—all far simpler than any Life form—only evolve as a result of our *ideas.* They *never* arise as the only seemingly clever results of the parts being randomly tossed about, inside and out, while we wait to see what wonderful new gadgets pop out, again and again.

So, did evolution, CHEMICALS-to-microbes, Microbes-to-Fish, Fish-to-Frogs, Dinosaurs-to-Birds, Apes-to-Men, take place?, *randomly,* as Darwin proposed, or *intelligently,* as his *spiritually* inclined followers argue? Or, did Life unfold in different ways? Our answer has to be that, given the Cambrian Explosion, given the *sudden,* not gradual appearances of the species, given the long-term *stability* of the species, we cannot argue that evolution, *as he described it,* took place at all.

Change over time? Yes. Microbes, Fish, Frogs, Dinosaurs, Birds, Mammals all happened. Change over time is a *fact*. But those changes, amazing and mysterious as they are do not support Darwin's ideas.

A far *better* hypothesis, than Darwin's, is that, at *every moment, Living Nature was—and it is—in-form-ed* by the *same, mysterious, underlying, second!, by second!, 24/7!, now!, now!, and now!, intelligence* that in-forms everything, including your amazing *Living Body and mine*, my mysterious *Sentiency* and yours, *your amazing Mind and everyone's,* our mysterious *Breath, and our astonishing Lungs, our amazing Brains, our Livers and Kidneys, our wonderful Muscles, our unceasingly beating Hearts and ever circling Blood.*

See a Spider weaving its fantastic, sticky, web. Observe the gyring Birds and gentle Butterflies. If we do not realize these amazing things to be *intelligent and meaningful expressions of intelligent and meaningful forces,* we no longer know how to look. It means that we have lost contact with our *Hearts*—and with the *Heart and Soul of things.*

To make matters worse, adding further to the muddle, we have the mistaken conflation of two other forms of evolution: micro-evolution, which is not controversial, *versus* Darwin's theory of macro-evolution, which *is* controversial, and much contested.

Micro-evolution describes minor variations *within* species—like finch beak changes. Macro-evolution is Darwin's thesis that all living things descended from microbes. Only, there's *no data* for such a process. Darwin's idea is, now, like the *Emperor's New Clothes* story.

Those who claimed to see the Emperor's amazing new clothes, were also busy explaining *how they were made.* Similarly for Darwin's theory. Those who continue to insist that it looks magnificent claim that its clothes are woven from random genetic copying errors!

Random genetic typos, we are to agree, drove Darwin's process of random microbes morphing into ra*nd0*m Worms, ra*n*dom Snails, r*a*n*d*om Fish, and, finally, ran*d*om you and random me. *Rand0m genetic* *typos* spun the *non-existent gold* of his ideas into the web spinning Spider and the gyring Falcon, the Salmon hunting Bear and the mighty *Whale!* Isn't it obvious? Maybe. . . if we don't think it through . . .

Darwin didn't know about genes, but this version of his theory is called neo-Darwinism. Neo-Darwinism proposes *random* genetic copying mistakes, in Life's complex DNA Coding, as the mechanism to drive his hypothetical process of the descent of all species—bacteria-to-Bears, Bears-to-Whales—with MASSIVE changes all along the way.

This fantastical process for which there is *no good evidence.*

Unfortunately, most modern *'evolution'* thinking is really just an *alternative creation story* for those who wish to deny the *spiritual.* But because it is not the kind of *practical science* that *has* to be right, or a plane will crash, or a drug will kill, a lot of us are not that bothered.

In this area of science, we can (just about) make up any story we like, then, proclaim: *"Don't worry,* we have it all worked out…you, your *ideas, your loves, and your dreams,* are just the *clever-seeming* products of *random genetic copying errors.* Don't bother thinking it through, trust us, Life is just an ACCIDENT and Darwin proved it."

Forgive the parody, but that's the size of it. *We just need to believe!* Random genetic typos, turned *single-cell* microbes into vastly LARGER *multi-cellular* organisms like worms, snails, and fish. Random genetic typos (RGTs) turned water-breathing fish into air-breathing amphibians. RGTs turned dinosaurs into birds, with *very* different lungs again.

Millions of rare, (spot the contradiction), RGTs transformed burly bears, or some other mammal, into echo-locating, deep-sea-diving whales. Random genetic typos turned rats into fast-flying, sonar equipped bats! RGTs turned caterpillars into beautiful butterflies . . .

Only, there is NO, GOOD, EMPIRICAL, EVIDENCE for *any* of these imagined processes. Darwin had an interesting idea. But, was it true? He had noticed that by *careful, clever* selection, *thoughtful, clever* people could bring about some *limited* changes to *existing* species, like wolves to dogs or pigeon or chicken breeding.

Some call these artificially induced—*within* the species—changes, like wolves to dogs, or minor finch beak variations in wild nature, ***micro-evolution***. But Darwin made no allowance for the fact that we, very clever humans, have only ever been able to *change a species* up to certain *genetic limits* which we have *never* been able to cross without *causing damage*. There is, however, no evidence that micro-evolution, *artificially guided by us,* or *natural,* has ever led to macro-evolution, to a new type of plant or animal. But, this was Darwin's CORE idea. The species ***had*** *to vary, massively,* for his theory to be true. Microbe to worm—*HUGE change!* Microbe to snail—*HUGE change*. Microbe to fish—*enormous change*. Fish to amphibian—*massive change*.

Yet, repeating, we, *focused, clever, humans,* despite *all* our efforts, have *never* been able to change any existing species beyond certain genetic limits without causing damage, let alone to create a new species.

No massive changes, then, when it comes to our—*determined, intelligently designed*—efforts to create new species. Yet Darwin and his fans have always *ignored* these key points and made unjustified extrapolations from micro- to macro-evolution. Unjustified in theory, because Darwin had NO EVIDENCE that such massive changes *might* be possible, and, unjustified in practice, because there are NO FOSSILIZED—*or living—transitional species* ever to be found.

This, given the *strict limits* to the changes which can be brought about by artificial breeding, is not surprising. It has *never* been possible to *artificially* evolve a species, beyond the limits set by its gene pool—without damaging it. Wolves to Chihuahuas? Yes. Wolves to a new species? No.

Darwin's idea was that if an existing species, like a Wolf, gave birth to a different version of itself, like a Chihuahua, it might survive and become a new species. But what he called *natural selection* tends to *confirm,* not to deviate species, let alone to create completely new ones.

Natural selection, it is clear by now, is *not* the highly creative new *natural law* Darwin thought he had found. Far from being an incredibly powerful creative force, natural selection has *no reason* to select any chemicals in the first place, let alone to turn them into Algae or Bacteria, Flowers or Fish, never mind Bears, Whales, Eagles or People.

Darwin's scheme can only make sense to us if we *truly* believe that there is *no intelligence or purpose* to the workings of *Living things,* that *Nature is Mindless,* and that the only 'creative' force that can exist is PURE CHANCE: Those like chemist, George Wald, who write,

"the spontaneous generation of a living organism is ***impossible,"*** then add, "Yet we are here as a result, I believe, of spontaneous generation."[117] Why? Because, "we choose to *believe* the impossible," *not* on scientific grounds, but on *"philosophical* grounds."[118]

Those like the biologist, Richard Lewontin, who write,

"Our willingness to accept scientific claims that are against common sense is the key to an understanding of the real struggle between science [which he equates with materialism] and the *supernatural* [which he equates with superstition]. ... Moreover that MATERIALISM is absolute, for we cannot allow a Divine Foot [intelligent causation] in the door."[119]

Unfortunately, this *[un]scientific* clinging to Darwin distorts science. Because all his 'law' of natural selection can do is to allow the *existing* members of a species, best adapted to the current conditions, to breed more prolifically than the less well adapted ones. In conflict with Darwin's predictions, it does not *deviate* them to any great degree.

Feeble or unviable variations of the original kind, like Chihuahuas, will not survive in the wild, and healthy variations will be even more wolflike than before—faster, and cleverer. Yet, as Michael Flannery writes, in *'**Was Darwin a Scholar or a Pitchman (skilled salesman)**?,'*

> 'For Darwin, the fact that man could breed a fancy pigeon or an especially fast race horse or a unique dog indicated evolution "in action." But, as [his colleague, Alfred] Wallace pointed out to him, when left in the wild, these fancy breeds [like Chihuahuas] *either perish or revert* to their original type.
>
> 'Besides, domestic breeding of animals requires the very thing Darwin sought to avoid — *careful thought* and pre-selection. In effect, it requires a breeding plan and design. This

> is clearly *not random* and purposeless, wholly natural causes operating to produce speciation. Darwin never saw that logical flaw in his own theory, yet it was obvious to naturalist Alfred Wallace, zoologist Pierre Grassé, historian Jacques Barzun, and many others.' [120]

Darwin, regrettably, was *unwilling* to accept the implications of some of the most obvious flaws in his ideas. The fossils *never* supported his claims for such a vast and 'indefinite departure,'[121] as from microbes to worms, or snails, or spiders, or bears, or whales, or, ultimately, you and me.

We also now know that cells use incredibly sophisticated means to, "protect themselves against precisely the kinds of random genetic change that, according to conventional [Darwinian] theory, are the [very] sources of evolutionary variability."[122]

All our, human, attempts to breed or to evolve any species beyond a certain point have *never resulted* in a new species arising, but only in existing ones being *damaged*. Domestic dogs damaged by overbreeding and fruit flies deformed in lab experiments are well known examples.

It is true that in the wild, when conditions vary, minor changes can take place *within* species, due to the potential for *some* adaptive variability built into their gene pools.

If the bark on trees becomes darker, as happened during the industrial revolution, the darker moths, *already in existence,* do better than the lighter ones which get eaten more often than the darker ones.

But when the pollution ends, and the tree barks become lighter, the fairer moths, which never completely disappeared, begin to predominate again. These kinds of *minor,* backwards and forwards shifts, have, falsely, been claimed as evidence for Darwinian macro-evolution.

Falsely, because these changes do not depend on RGTs. They just take advantage of the moths' *existing* genes, arising from an *existing* gene pool which *already includes* possibilities for darker and lighter examples to hatch, or finches with larger or smaller beaks to hatch, the successfully surviving proportions of which vary as conditions change.

But are not the horse fossils and archaeopteryx evidence for evolution? Well, in ***The Naked Emperor: Darwinism Exposed,*** Antony Latham argues that the horse fossils point to micro-evolution, not to macro. He writes, "We do not see, in the horse series, any major change in anatomy, just loss of digits, increased size and altered tooth shape. This hardly constitutes evolution on the grand [macro] scale." [123]

Jonathan Wells, in his book, ***Icons of Evolution: Science or Myth?,*** tells us that Henry Gee of *Nature* magazine, while remaining loyal to Darwin. overall, "candidly admits "that we can't infer descent with modification [for horses or any other species] from fossils." [124]

"No fossil is buried with its birth certificate ... To take a line of fossils [like equines] and claim that they represent a lineage is *not* a scientific hypothesis that can be tested but an *assertion* that carries the same validity as a bedtime story – amusing, perhaps even instructive, but *not scientific."*[125]

The primary issue, though, is *not* succession but *cause*. We create, and evolve, things in *successions* all the time. Modern Fords are the conceptual descendants of earlier Fords. But, every Ford, since T, is an *intelligently* created successor. It did not evolve by pure *luck!*

So, even if Darwin was right about macro-evolution, microbe to fish, spider or snail, it would have needed to be an *intelligently* driven process. But there is *no evidence* for his ***key*** idea that one species can slowly turn into a completely different type—be it randomly as his *atheistic* fans assert, or *intelligently* as his *spiritually* minded followers claim. As John Davidson explains in ***Natural Creation or Natural Selection,*** "NOWHERE has any fossilized species been found which is *on the way* to becoming a *Spider [or Fish or Bear or Whale or Bat or Butterfly*]. It is spiders [or fish or bears or bats] or something else!" [126]

Are we, then, to ignore all the Darwin-falsifying data, just to try to make his steam-age yarn seem *'right?'* That's not how science was supposed to be. As to Archaeopteryx, in *Icons of Evolution,* Jonathan Wells explains, that there are just *"too many structural differences* between Archaeopteryx and modern birds for the latter to be descendants of the former." He continues,

"In 1985 University of Kansas paleontologist Larry Martin wrote: "Archaeopteryx is *not ancestral* of any group of modern birds." Instead it is "the earliest member of a totally extinct group of birds."" [127] So, if archaeopteryx is not a credible ancestor for modern birds, is she a good candidate to be a descendent of earlier dinosaurs?

In *Icons of Evolution: Science or Myth?,* Wells discusses one theory which is that archaeopteryx *descended* from a group of dinosaurs the only fossil data for which comes in a stratigraphic period *long after* archaeopteryx was extinct.

But how, you may be thinking, could *later* dinosaurs be ancestors to a *much more ancient* animal which was *already long extinct* when they were around? It doesn't make sense. Wells is also skeptical: "The claim that birds are dinosaurs, strikes most people—including many biologists—as rather strange. ... it defies common sense. Birds and dinosaurs may be similar in some respects, but they are also very different."[128]

12 Darwin's Mistake

Darwin believed that species could *vary indefinitely*. In *Darwin's House of Cards,* Tom Bethel calls this "DARWIN'S MISTAKE." Yes, small differences can arise between the generations, but, he points out,

> "Darwin's mistake was to assume that these [minor, reversible] differences somehow accumulate over the millennia, so that one species eventually transforms itself into another. WITHOUT EVIDENCE Darwin's supporters today *still* [believe this]. ... BUT SUCH A TRANSFORMATION HAS NEVER BEEN OBSERVED [either in *living history* or in the fossil records]. *No species [fossilized or still existing] has ever been seen to evolve* into another."[129] Yet earlier species 'evolving,' repeatedly, into other, later species was Darwin's core idea. So why has Darwin's theory not yet died?

His theory has not yet died, not because it's true, but because its supporters have been unwilling to take it off life support. It is ironic too that although Darwin called his book, *"On the Origin of Species,"* he did not, in honest fact, explain the *Origin of Species* at all.

He had nothing *publicly* to say on the subject. His public stance was simply that, although we may not know how Life started, once it had done so, he, Darwin, had a great theory as to how it unfolded from there. The idea we now call Evolution. So, however it first arose, there was some *microbial Life,* like algae or bacteria… and some of these bacterial, or algal, cells *randomly varied,* and some of those *variants* survived, because they were beneficial, and this *led to new species* gradually forming—like fish, snails, spiders, or worms—not due to any underlying *intelligence*, but due to three new laws called (1) Natural Selection, of the random variations, (2) Survival of the Fittest random variations and (3) Descent, with VAST modifications, of the original *ancestral microbes* into fish, spiders, snails, whales etc.

What doesn't make sense, though, about Darwin's theory, is that such tiny algal or bacterial microbes would, already, have been *highly viable* organisms in their own right, *already surviving very well*, already *reproducing in massive numbers* as bacteria and algae do. So *why* why would they ever *need* to 'evolve' beyond such a stage, let alone do so?

They would have no 'natural' need whatsoever to become vastly more complex Life Forms. Vastly more complex Life Forms which would find it ***harder to survive,*** than their ancestor bacteria—*not easier!*

Bacteria are, arguably, the most successful Life Form, with no need, nor any realistic chance of evolving into anything else, let alone into VASTLY LARGER organisms like spiders, snails, bears, birds, bats or whales, all of which would find it much *harder to survive* than Bacteria.

Yet, according to Darwin's theory, such a microbe *might randomly vary* enough as, firstly, to acquire nuclei, which bacteria do not have, then to randomly clump together to become one of the first, *multi-celled* organisms, like a worm, mollusk, jelly fish, spider or fish by

> "differences [or variations] not greater than we see between the natural [wolves] and domestic varieties [dogs] of the same species at the present day. ... [where] the number of intermediate and transitional links, between all living and extinct species, must have been INCONCEIVABLY GREAT. But assuredly, if this theory be true, such [huge numbers of transitional species must surely] have lived upon the earth."[130]

This requires us to imagine each *random variation* gradually making its original *random-chemical-mess* of an unintended and 'purposeless' Life-Form into something truly beautiful and truly amazing, like an Ammonite, Trilobite, Star Fish, Worm, Spider, Beetle, Fish or Snail. Let us, *please!,* think: tiny bacterium turning into a VASTLY larger SNAIL.

A Snail is 10,000 times longer than a bacterium. Yet, according to Darwin's theory—which we are all supposed to believe in, and be thought stupid if we do not—a bacterium could evolve, by PURE CHANCE!, for no meaningful reason at all, into something so VASTLY LARGER as a SNAIL, or FISH, the largest of which, like a Whale Shark, is up to 20,000,000 million times longer than a bacterium.

With no offense to those who still believe in it, Darwin's theory does not work. However, our firm and clear claims here that Darwinian Evolution COULD NOT happen is not the same is it DID NOT happen.

We can say it COULD NOT happen, as much as we like, but, if it DID happen, we will be the ones who are wrong. Fine. But is there any ***evidence*** for such a process, as Darwin described? Of bacteria gradually, turning into snails, spiders or fish, let alone whales? No, there is none.

Darwin provided no data at all, let alone any *extraordinary* data, to back up his own quite *extraordinary* beliefs in the creative powers of aimless, random, causation—nor has anyone since.

As Michael Denton writes, "To common sense it seems incredible to attribute such ends [as *Cells, Blood, Hearts!, Livers, Eyes!, Brains*] to ***random search mechanisms,*** known by experience to be incapable, at least in finite time, of achieving even the simplest of ends.' [131] [*Ah, yes.*]

Denton reminds us that even Darwin had his own doubts about the unrealistic nature of his *scientifically untethered* claims. He wrote:

> "Although the belief that an organ so perfect as the eye could have been formed by [aimless/mindless] natural selection, is enough to stagger anyone … [yes] I have felt the difficulty far too keenly to be surprised at others hesitating to extend the principle of natural selection to so startling a length."[132]

One *scientific* way, to have tried to answer his own doubts, would have been to provide proper *probability estimates* to show that the paths to such complex mechanisms—as *Cells, Blood, Taste, Smell, Sonar, Kidneys, Eyes, Ears, Brains, the Caterpillar-to-Butterfly enigma, the mysteries of male and female, of asexual and sexual reproduction, of testicles, ovaries, wombs, livers, hearts, lungs,* and countless other biological systems—could have been found, by **'PURE CHANCE'** in the time available. Yet, as Michael Denton writes,

> "*nowhere* in the *Origin [of Species]* is *any* attempt made to provide [any] quantitative support for the grand [Darwinian] claim of the all sufficiency of **CHANCE** [as a creative principle].
>
> "It is true that Darwin appealed on many instances in the *Origin* to the enormous periods of time available to the evolutionary search but, as is the case with any other random search procedure, time in itself tells us NOTHING of the PROBABILITY of achieving any sort of goal UNLESS the COMPLEXITY of the search can be *quantified.*"[133]

Interestingly, at the Wistar conference in 1966,

> 'Some mathematicians did try to make the calculations. The mathematician D.S. Ulam argued that it was *highly improbable* that the eye could have evolved by the accumulation of small [rare, harmful, random, genetic] mutations, because the number of [random] mutations would have to be so large and the *time available was not nearly long enough* for them to appear.
>
> 'Sir Peter Medawar and C.H. Waddington responded that Ulam was doing his ***science backwards****[!];* the fact was that the ***eye had evolved****[!]* and therefore the mathematical difficulties must only be apparent[!]. Ernst Mayer observed that Ulam's calculations were based on assumptions that might be unfounded [while not questioning his own assumptions!], and concluded that "somehow or other[!] by ***adjusting these figures****[!]* we will come out all right[!]. We are comforted by the fact that evolution has occurred.[?]" [134] But, has evolution occurred? Saying 'eyes' or 'hearts' or 'brains' exist, therefore, they *'must have evolved'* is question begging. It assumes, in advance, the very thing which it needs to prove to us.

Yet, despite the mathematicians working out that Darwinism could not account for the eye, because their views conflicted with Darwin, it was they who *'had'* to be wrong, not Darwin! Darwin's entire theory is based on BLIND LUCK as its supreme creative principle. Yet, there is *no evidence,* from any time or place, that any *purposeful* mechanism can rise from ***ra*N**D**o*m*** causes—not even a simple wooden toy.

Reminder: if a bacterium were the size of a pea, the *smallest* Snail would be 50 meters long, and the largest Fish would have to be about 60 miles long! Around 20 million times longer! Is this Darwinian idea realistic? COULD it happen? Is there any evidence that it DID happen?

Such a stupendously complex biochemical process, if it happened due to some *amazingly intelligent—spiritually guiding—movement,* would be astonishing enough. But Darwin's idea that *such, already highly viable* bacterial cells would behave in these staggeringly complicated ways—for no reason at all—is beyond unrealistic.

But, realistic or not, this is the essence of Darwin's theory. He called it *descent with modification.* We, humans, are, according to Darwin's thinking, *descended, by UNINTELLIGENT CHANCE,* from microbes.

Why argue, though, for such an *impossibly unlikely* process? Because, ***philosophically***, Darwin, like many of his fans since he, didn't like the idea of an *intelligently or spiritually driven creation.*

He could not, it seems, get his head around the classical idea that we are born of an *Intelligent Spiritual Mystery*, which ***just Is,*** therefore we had to be born from a MINDLESS mystery, which ***just is***.

An *unintelligent creation,* driven by pure chance, by *'unintelligent design,'* had to be his alternative. The trouble is, there is so much *ideology* behind Darwin's way of thinking.

He could not believe that the 'good' god of the Christian teaching with which he was familiar, could *deliberately* create so many horrible creatures, so many diseases, and so much suffering. So, perhaps, there was no creation at all. He was not familiar with the more sophisticated

Hindu concept of one Ultimate God called *Brahman,* like *God the Father* in Christianity, *who, or which,* ***Just Is,*** but which, in Creation, *the Son,* takes three very different forms: Brahma the *Spirit of Creation.* Vishnu the *Spirit that Sustains and Maintains* Creation, moment by moment, and Shiva, the *Spirit of Death and Dissolution,* a Force without which nothing created could ever be dissolved, in order to allow the new to emerge.

It is partly Darwin's, and others, childish demands that Creation be exactly as we, tiny, puny, humans would like it to be, which has caused so many modern thinkers to conclude that there cannot be any *God or Spirit or Intelligent Creative* forces at all. Why not? Above all because we cannot seem them *physically,* and because we wish to criticize this amazing creation, despite it *sustaining us, right now!, now!, and now!, this amazing second!, and every amazing! second, 24/7.*

Personally, I hope, and trust, that the *Ultimate, Spiritual, Creative Force(s)* of our Universe are not just ***intelligent*** but also ***benevolent.***

But our arguments, in this book, are not, primarily, about the benevolence, or otherwise, of *Spirit or God or the Tao or Universal Creative Spiritual Beingness, or your preferred label,* they are about the amazing *intelligence and purposefulness* of those *Creative Forces.*

We need not disagree with Darwin that the problems of 'natural evil,' like predation, disease, and suffering are hard to understand. But the idea that INANIMATE CHEMICALS would become *Alive!,* by pure chance, for no reason at all, is far more illogical than the pre-Darwinian idea that the Universe, and we in it, is the product of *stupendously intelligent forces, which* ***just Are***—not the totally mindless ones of Darwin's materialistic beliefs, which, no less *mysteriously,* ***just are***.

But, this was his view. The species did not arise due to any *inherent intelligence* but they had gradually descended, due to *random variations* from microbes, into VASTLY LARGER plants or creatures of many different kinds, in the shape of an ever spreading *Tree of Evolution.*

Let us imagine, though, the vast numbers, the millions upon millions of *random* biochemical changes, which would need to happen for a bacterium to gradually *evolve* into a 10,000 x larger SNAIL with a HARD SHELL, or a WEB-SPINNING Spider, with eight legs, or a Spineless Worm, or a 20,000,000 times larger Fish with a Backbone?

Tiny bacterium, evolving, BY CHANCE!, into Trilobite, Spider, Snail, Worm or Fish? *HOW?* Fish evolving, due to millions of *'random'* biochemical changes into amphibians with *very different lungs, HOW?*

Dinosaurs evolving into birds with *very different lungs? HOW?*

By pure chance? Seriously? Although Darwin, misleadingly, called his book ***On the Origin of Species,*** he did not tackle the origins of *Life* at all. Yet, with *no scientific* justification, what started, in Darwin's day, as mere speculation, has become a dogma of modern science:

"Life is nothing truly *miraculous or supernatural.* No, actually,

"You, your dear child or beloved pet may seem *magical and meaningful,* but really all you, or they, are is *just a bunch of chemicals,* and, when the chemicals stop, you drop and that's all you ever were."

It *sounds* irrational dear reader, and it ***is*** irrational, deeply irrational, and it is shockingly un-scientific, but we need to take into account this strange modern way of thinking, because it lies behind all our contemporary arguments about Life's origins and its true history.

Not all scientists are *materialists,* like Darwin, but, today, at the heights of science, the **MATERIALIST** view is the *dominant* view. 90% + of scientists in America's most prestigious scientific body, the National Academy of Sciences, the NAS, do profess to *atheism.*[135]

This is why those few scientists currently *brave enough* to argue for intelligent design or ID are ignored or ridiculed. One criticism being to accuse ID of being religiously motivated. Well, we can, just as well, argue that Darwin and many of his fans are *atheistically* motivated.

They don't believe in the *supernatural,* so they insist on blind-chance theories of Life's origins, while maintaining that they are the *true* defenders of science. It would be amusing if it wasn't so damaging to us and to our relationships to the *deeper, spiritual* sources of things.

The *scientific* approach, the *rational* approach, they say, is to argue that *all Living things* arose, and evolved from there, by *pure, dull, chance!* Why, however, may not other scientists disagree with this?, as they do, and falsify it?, as they have done—repeatedly. Especially, as, until Darwin came along, to argue that, "Living things *"look"* intelligently designed, because *they really are!,"* was the classical view.

Let us, again, try to imagine a Cell, DNA Code, Blood, Heart, or Brain arising by PURE CHANCE. Will vast time do it? No. It is easy to show that it won't. Blind chance has no reason to make anything at all.

Remember, "it is MORE LIKELY that you *and your entire extended family* would win the state lottery EVERY WEEK for a MILLION YEARS than for a bacterium to form by chance!" [136]

Yet, nowadays, we have to plead for what, before Darwin, all the greatest thinkers believed: from Plato to Galileo, from Pythagoras to Planck, from Newton to Maxwell, they all believed in intelligent design.

Yet so many of us now claim that we *know* for a fact that Life arose by *pure, mindless, chance* and that to introduce any ideas of the *spiritual* into our understandings of Life is ignorant and unscientific!

This *deny-the-spiritual-at-all-costs,* approach in modern science has distorted our relationship to each other, to ourselves, and to creation as a whole. I understand that, for many of us, this may partly be because we would be rather be *free!,* than be told what to believe, by this or that religious authority. Of course, I get it. One of my friends once said to me, "If I believed, as you do, in the *spiritual,* I'd have to become a Christian, and I'm not sure if I want to do that…"

I replied, "Look, I'm not asking you to *join* any religion, let alone to become a *Christian,* although there's so much to admire (as well as to criticize) in that vast tradition, and its many different forms. No, just to consider that there may be more to reality overall than is dreamed of in your *spirit-denying* philosophy. That reality is *multi-dimensional,* not PHYSICAL and MATERIAL alone, and it may be far more magical and meaningful than you have allowed yourself to believe, that's all…"

13 Unintelligent Design?

In ***Signature in the Cell: DNA and the Evidence for Intelligent Design,*** Stephen Meyer explains how scientists are supposed to search for the best and most *causally adequate* explanation for a given phenomenon.

So, is mere CHANCE *causally adequate* to create anything clever? Or is *intelligence,* be it *subtle* or be it obvious, the more logical and more rational explanation?

Prior to Darwin, *spiritual intelligence* was assumed to be the cause of Living things. Then, some, like he, began to argue that pure luck could create *Life.* However, in ***Darwin's Black Box,*** Michael Behe explains that mere luck *cannot* give rise to mechanisms which are *irreducibly* complex: mechanisms that cannot emerge in a step by step process of gradually accumulating complexity, as Darwin proposed.

For example, the cell's very clever—*selectively permeable*—outer wall, CODED for by the very DNA, *within* the cell, which the cell's outer wall protects. They *must arrive together,* or not at all.

Avian lungs: different to those of all other vertebrates. They could not evolve gradually, without causing death, more or less instantly.

The *bacterial flagellum*: the tiny tail bacteria propel themselves with, powered by a nano-scale motor which looks remarkably like our own electric motors, incredibly machine-like in fact.

Professor Behe explains that this amazing nano-scale machine could *not* arise due to a series of random, Darwinian-style processes.

Why not? Because, just like the motors we make, *all* the parts would need to be there at the time of assembly, or it would not work at all. This is why Behe's hypothesis is that only an in.FORM.ing *intelligence* of some kind, could give rise to this clever bacterial propeller, not BLIND CHANCE. Because there is no adaptive advantage to just one or two of the flagellum's parts being there, so they would *never* be *'naturally'* selected. No, just like the motors we make, the flagellar motor is a product of an *underlying intelligence* inherent to the overall *spiritual reality* of which our *own intelligences* are parts. Why so many of us continue to argue, in the name of *'science,'* that *'pure luck'* can create anything remotely clever is a mystery to me. *How is that scientific?*

"We cannot *physically* see Life's *deeper, intelligent,* sources, so mere LUCK *must* create all..." seems to be the only semi-logical reply.

Now, some of those who cannot see, what seems so obvious, I hope, to you and to me, have tried to rebut Michael Behe by arguing:

"This or that part of the bacterial motor *might* have been part of something else, and, it *might* have become a part for the flagellum, and this *might* disprove Dr Behe's theory." But this just pours more unlikely speculation onto more unlikely speculation. *'Might this... might that...'*

In ***No Free Lunch: Why Specified Complexity Cannot Be Purchased without Intelligence,*** William Dembski explains the [utterly] obvious point that mere chance is *incapable* of generating the totally specific sequences of parts and software which define all biological mechanisms and, indeed, all the mechanisms we make.

Our materialist-atheist thinking friends, like Charles Darwin, Richard Dawkins and Richard Lewontin, disagree, as, by virtue of their *anti-spiritualist* philosophy, they *have* to disagree. In fact, Dawkins, among others, has, with an irony which seems to escape them, *intelligently designed* software to try to show how living mechanisms could, they claim, evolve by random-no-intelligence-needed, processes.

William Dembski shows, however, that these computer scenarios all fail. Why? Because they all *smuggle in* pre-defined—i.e. *intelligently designed!*—search criteria which are neither truly random, nor *'unintelligent,'* nor available in the *'directionless'* and *'purposeless'* nature of MATERIALIST belief, while deluding themselves, and their audiences, that they do not. A reviewer of *No Free Lunch* writes,

> 'Dembski describes the 'shell game' [3 cups trick] whereby evolutionary programmers try to obscure the 'smuggled in' information and claim that their programs have generated the output[ted] information from scratch. ... My only hope is that

critics will have the decency to actually engage the ideas presented here. ... Dembski has ... 'taken on' the established Darwinian orthodoxy, demonstrating the insufficiency of non-intelligent processes to generate the specified [non-random] complexity [that defines all Living things].' [137]

As William Dembski says, neither regular laws alone, nor **CHANCE** alone nor some mix of the two, can generate the *totally specific* sequences of **DNA** and parts which make possible the biological mechanisms which Behe discusses.

In *Billions of Missing Links: A Rational Look at the Mysteries Evolution Can't Explain,* Geoffrey Simmons gives many examples of biological complexity which *cannot* be explained away, by references to *'evolution,'* or to *'lucky'* random mutations, or *'might this, or, might that... '* For example, the astonishing *defense techniques* of some insects:

> 'An Australian termite called the nasui ... If provoked by a predator, it will swing its head side to side and squirt loops of a chemical mix. One component *will glue* the attacker in place, another will *paralyze it,* another may kill it, and a fourth will attract other termite soldiers to gather around the intruder and to keep spraying it. Their sprays contain alpha-pinene, beta-pinene, limonene, trinervitrenes, and kempanes. The latter two chemicals cause the stickiness and they have *not been found* anywhere else in nature. This species also *defies* [Darwinian] *evolutionary logic.* [138]

'The bombardier beetle's weaponry is equally extraordinary.

> 'This African insect can fire off two chemicals, hydrogen peroxide and hydro quinone, *from separate storage tanks and rear jets.* [The beetle can fire accurately and repeatedly. While separate the compounds are harmless. When mixed they *instantly heat up* and hit their opponent at *boiling point,* 100°C!] When the chemicals combine, they form a new

chemical that burns the predator. The beetle can shoot these chemicals with an uncanny accuracy, as well, to either side, backward, or even forward, by swinging its tail under its abdomen. *Special nozzles* blast predators at a rate of *500 bursts per second,* each at a speed of 65 ft./s. These chemicals are potent enough to severely damage a mouse and injure the eyes of any animal . . . Yet these chemicals are entirely benign when stored separately at the back end of these beetles. *How could this happen by accident?* "Oops, those two chemicals didn't work" (spoken by an intermediate species).

"Mind if I try two others before you eat me?" Or, "Could you stand a little taller so I can get you with my nozzles?"

'Keep in mind there are hundreds of thousands of chemicals on this planet to choose from. And even if the combo turned out perfectly right the first time, the beetles still needed a way to make them, store them, and fire them off. [139]

'Are these all *merely lucky* chemical mutations that protected the species? ... How did the *special storage glands* that protect the owner evolve? Before or after? All at the same time? What happened to the thousands of *missing links?* [There is no fossil (or other) evidence for any such links.]' [140]

Simmons says of the Bombardier's defense system,

'The entire mechanism defies [Darwinian] evolution. *How* and *why* would there have been a *gradual development of chemicals* that were *useless* until they finally reached their present form? This system is *all or nothing.'* It is irreducible.

At the most dynamic level of all, at the *intracellular level, irreducible* complexity is pervasive. Cells, which we, *unlike* Darwin and his steam-era friends, now realize, are like vast, micro-miniaturized, chemical-factory-worlds, in tiny spaces *smaller than sand grains—think about that!*—contain nano-scale, bio-robot machines, called proteasomes. An article in *Nature magazine,* discussing these, notes,

"Correctly dismantling a structure can be as challenging as assembling it."[141] *Science and Culture Today* commented on this article.

> 'Thus begins an article in *Nature* about the cell's shredder-recycler, the proteasome. This… [clever] molecular machine ... validates the trash… and shreds it… Inside … [mechanisms] cleave polypeptide chains into short segments about 7–9 amino acid units long, which can be reused by the cell directly, or further degraded into individual amino acids for recycling. . . An elaborate lid structure … guards the entry gate and… [to] be validated, a target protein must have been previously tagged for destruction by *other molecular machines.* …
>
> 'Proteasomes move throughout the cytoplasm and the nucleus carrying on their vital work every day [24/7] inside your body, identifying and selecting proteins to degrade, keeping the Cell free of damaged or mis-folded proteins, and removing proteins whose work is done. ... These intricate machines help keep the immune system functioning and are ever-present to respond to stress. Failure of these systems can cause serious diseases, like Parkinson's and Alzheimer's, or death.'[142]

Science and Culture Today (formerly *Evolution News*) comments that although the "authors of the original paper said nothing about evolution, Geng Tian and Daniel Finley, in their summary of the paper in the magazine *Nature,* couldn't resist" introducing a bit of traditional, unfounded, speculative, Darwinian-style evolutionism. They wrote,

> "The eukaryotic proteasome seems to have evolved [HOW?] from a protease known as PAN … which is found in microorganisms called archaea." They suggested. Then they produced a colorful diagram called,
>
> "The evolution of proteases," showing trypsin, PAN, and the full-fledged proteasome [adding] "This dramatic evolutionary elaboration of the protein-degrading machinery is

> reflected in the fact that the proteasome has assumed regulatory functions in virtually all aspects of eukaryotic cell biology ... The evolution of ubiquitin tagging also coincided with a transformation of the proteasome's structure."

'*Science and Culture* asks, "Did they SHOW how [random] mutations and *unguided processes* produced this highly complex machine, let alone the other systems it interacts with? No; they just assumed it. *"The evolution of"* begs the question.

"Comparing three functional designs of varying complexity doesn't prove they evolved [one from the other] any more than showing a *scissors, a shredder, and a recycling system* proves they evolved from each other by undirected [mindless] processes."[143]

It is easy to claim—by virtue of a prior commitment to materialism —that this or that stunningly clever bio-mechanism *must have* "evolved" without bothering to ask, *"HOW* did it evolve? *Could* it evolve? *WHY* would it evolve?" As biochemist, Michael Behe, writes,

> 'There is *no publication* in the scientific literature in prestigious journals, specialty journals, or books that describes *how* molecular evolution of any real, complex, biochemical system either *did* occur or even *might* have occurred. There are [the usual] *assertions* that such evolution occurred, but *absolutely none* are supported by pertinent experiments or calculations.' [144]

Here's another interesting thing. How many of us now realize, or remember, if we once knew, that what we think of as *Darwin's* theory actually had *two* inventors, and it was, initially, thought of as belonging to both of them, not just to Darwin. The other being Alfred Russel Wallace whose ideas, initially, were very similar to Darwin's. But he, unlike Darwin, eventually worked out that no *pure-chance-based* theory of living things would ever work.

Interestingly, Wallace also carefully investigated *spiritualism* and *after-life* communications from those passed on,[145] and—[along with other 19th century luminaries, like Sir Arthur Conan Doyle (literary genius, *Sherlock Holmes*), Sir Oliver Lodge F.R.S. (world class scientist, spark plugs), Professor James Mapes (chemist), Frederick Myers (classicist and psychic researcher), Sir William Barrett F.R.S (physics), Professor Camille Flammarion (founder of the French Astronomical Society)]—he concluded, "that the phenomena of *Spiritualism*... do not require further confirmation. They are proved quite as well as facts are proved in other sciences.'[146]

Apart from the powerful, modern, *spiritualist* research[147] which falsified Darwin's belief system, there were many other things, Wallace realized, which his and Darwin's theory did not explain.

Like what? Like the amazing *metamorphosis* of the crawling caterpillar into the beautiful Butterfly.[148] Wallace worked out—not that it needed a genius to work this out—that there was *no way* mere chance could account for the stunningly complex transformations involved.

A caterpillar, he realized, is a viable organism in its own right. There's *no reason* for the rAnDom caterpillar of Darwinian thinking to *randomly* dissolve itself into a life-forceless[149] and *purposeless* chemical-fluke soup, then, *accidentally,* re-emerge as a beautiful Butterfly. HOW could a *random genetic typo* cause a caterpillar to mutate, a little, towards making a cocoon?

Even if a caterpillar did, by pure chance, start to make a cocoon, what would happen next? Let's imagine it, and think it through, so no one can accuse us of being intellectually lazy.

Firstly, cells, we now know, *protect against random genetic errors,* by very clever means. Such a dramatic change, as even just *starting* to form a cocoon, would require *millions* of mutations, not merely one or two. So the process would never even get started.

Then, even if such an unlikely thing happened, *why* would it be naturally selected? It would confer no survival advantage, let alone lead to a beautiful Butterfly, but only to *death!* There'd be nothing to cause the ex-caterpillar to return to its prior condition, let alone forward, to a never-before-seen butterfly. For any caterpillar starting to mutate, *only death* would await. Dear Reader, as Wallace realized, his and Darwin's theorizing, ultimately, didn't work, it didn't make sense.

There is *no survival* advantage in merely *starting* to form a cocoon. What, after all, would cause the, 'it-used-to-be-a-caterpillar,' now, 'it's-a-useless-soup-in-a-cocoon,' to morph into a beautiful Butterfly, necessitating *millions* of complex biochemical changes, not just one or two? The answer is… absolutely nothing! No blind process could cause the *'purposeless soup'*[150] the Caterpillar had dissolved into, to turn into a magnificent Butterfly. The idea that *random genetic typos* could give rise to the miraculous bio.logical happening that is a Caterpillar, let alone to a beautiful Butterfly, is unreal. Bio = *Life!* Logos = *the Word!,* the *Soul* of in-FORM-ation—to come into form.

Darwin had observed that *skillful breeders* could modify *existing* species—like tulips, pigeons, or dogs—to a *limited* degree. But to extrapolate from this, as he did, that one species could change into a wholly other species—via a shared descent—with HUGE changes, all along the way, due to *rare, genetic copying mistakes,* was unjustified, then or now. He, not having our modern knowledge, did not realize that Life could never arise by mere luck. It was impossible.

Had Darwin known these things, he would, surely, have conceded,

> 'I realize, today, that the idea I mentioned to my friend Hooker, that Life might have arisen, by mere CHANCE, in a warm little pond somewhere, is impossible. It is obvious to me, now, that such transformations as the caterpillar-to-butterfly process, *could not possibly* "have been formed by *numerous, successive, slight modifications.*" This means that *my theory breaks down.* To try to explain all of Life as just a very long series of *extremely* [impossibly] *unlikely* chemical ACCIDENTS no longer makes sense.
>
> 'In light of what we now know, we have to admit that there is a *tremendous intelligence,* to the 24/7 functioning of Living things. It would be *unscientific* to deny it. Our ancestors were

> right, *intelligence and intelligent causation, must be* parts of the basic fabric of reality. The idea that *Being-Existence, that What Truly Is,* is MINDLESS to its core, the idea that *bright, intelligent minds,* like yours and mine, could arise by pure, *mindless* LUCK, no longer makes *any scientific* sense—if, indeed, it *ever* made sense.This is why, if we are to make any further *scientific* progress on *Life's* origins, we must accept this, even if it feels like a blow to the pride of of our prior commitments to MATERIALISM and to aimless, unintelligent, DULL, CHANCE, as a supreme creative principle.'

Darwin wrote, in *The Origin of Species,*

> 'If it could be demonstrated that any complex organ existed, [*Cell, Blood, Brain, Eye, Heart, Lung*] which *could not possibly* have been formed by numerous, successive, slight modifications, my theory would *absolutely break down.* But I can find out no such case.' [Ah, Dear Darwin, it is not hard to find thousands, if you look…]

In the light of our knowledge today, it would make as much sense for Darwin's intellectual descendants to confidently proclaim,

> "If it could be demonstrated that any complex, AI guided Robot, Car, TV or Computer, [all simpler than any Living thing], could not possibly have been formed by numerous, successive, slight modifications, driven by nothing more than the laws of *random mixing,* in combination with nearly infinite amounts of blind LUCK, like you and your family winning "the state lottery EVERY WEEK for a MILLION YEARS!"[151] our theory of evolution by pure dull chance (PDC) *would absolutely break down.* But we can find out no such case!"

Such a statement would be absurd. But, when it comes to living things, the usual rules no longer seem to apply. It is quite clear, by now, that there is nothing in the laws of chemistry—unguided—to cause living things to arise.

In 1859, Darwin put forward a new theory: all of Life on Earth, had descended from just one, or two, universal ancestor microbes, with *HUGE modifications,* all the way, *if,* such a microbe could vary almost *infinitely!* Microbes to Men—due to totally *mindless* processes.

A theory which the late 20th century materialist philosopher, Daniel Dennett referred to in the title of one of his books as ***Darwin's Dangerous Idea***. Why? Because Darwin had questioned so many previous cultural certainties.

Given the physical invisibility of Life's deeper sources, rather than continuing to acknowledge all the obvious evidence for *intelligence and purposefulness* in nature, and take the classical, *spiritualist* view, of thinkers like Plato and Galileo, Pythagoras and Planck, Darwin wondered whether Life *had* any deeper sources?

"What if," he asked, "there was no *Intelligent Cause of Anything?" "What if,* long ago, there was just a warm little pond somewhere, with some CHEMICALS in it, some flashes of lightning, and, *voilà!, Amazingly, Astonishingly, Staggeringly, Complex Life!"*

Darwin did not know enough biochemistry to understand that this was *a chemical impossibility* and, as Eastman and Missler note, a mathematical absurdity. Yet, because of his commitment to materialism, to try to make his own forever unproven views on origins seem true, Darwin had to disregard all our *real-world* experience, scientific and everyday which tells us that no complex mechanism ever *'falls together'* by *probability-violating, entropy-defying, chance, by Unintelligent (non) Design.*

Of course, the *real danger* in Darwin's idea, unnoticed by Dennett, and so many others, in the years since Darwin, was in the *harm* it would eventually do to *science* and to very *rationality* itself. The pretense that there is no *intelligence or purpose* to how living things are made, or function, *24/7,* is so clearly untrue. It is, also, clearly not true that 'MERE LUCK' can create anything remotely functional.

Unfortunately, after Darwin, the denial of *any* intelligence or purpose, *at all,* in the making, and *second, by second, 24/7 functioning,* of Living things and of the *spiritual* in general, in the name of *science, did make* our science culture, in the area of origins, *irrational.*

It led to a breakdown in *spiritual* belief, and left us with a science adept at *measuring* and *using* many things in Nature, but hopeless at seeing the *true* values and the *inner* qualities—*the Soul*—of anything.

It is very clear by now, even allowing for the PHYSICAL invisibility of Life's *deeper* sources, that MATERIALISM does not make the best sense of the data overall.

Yet, for this sterile philosophy to continue to seem *true,* so much reason, logic, intuition, and, for those who have them, *psychic, spiritual, and mystical* experiences, have to be dismissed or denied, because, if this is not done, materialism *is falsified.* We, though, who know so much more than Darwin and his early supporters, can no longer justify thinking as they did. If, despite this, we continue to do so, it must be, as Hoyle said, for reasons of *emotion,* not ones of science.

Yet any dislike, we may have, for the religions, or for their officials, is not a good enough reason for us to continue to pretend that we *truly know* what Life is, than to admit that, really, we still don't have a clue.

No, sadly, so much of the Darwinian evolution story is now about clinging to an explanation which seemed plausible [to some] at the time, then refusing to accept that his, and his supporters ideas, although they initially seemed interesting, did not hold up. Despite the premature claims of his admirers, despite calling his book *On the Origin of Species,* Darwin did not explain the origins of Life—at all.

Not only this, but we have, today, not just one but *many* demonstrations of the kind which show that *Darwin's theory has "absolutely broken down."* Here are just a few. There are many more, and, with just a little reflection, we can, all of us, find plenty more examples of our own.

1. Life,[152] and Living Cells, the mind-boggling complexities of which Darwin and his Victorian peers knew vastly less than we.

2. *Irreducible Mind. Immaterial, Intelligent, Curious, Creative Mind* cannot be reduced to MINDLESS MATTER.[153] It's obvious.
3. The *irreducibly* complex bacterial flagellum and its nano-scale motor,[154] of which Darwin and his contemporaries knew nothing.
4. The *irreducibly* complex biochemistries of vision and of blood clotting discussed by Behe,[155] of which Darwin knew nothing.
5. The *irreducibly* complex defense technologies of insects discussed by Simmons in his book *Billions of Missing Links.*
6. The *metamorphosis* of the Butterfly considered by Wallace.
7. Life's *stunningly complex* DNA CODING which, like all code, could *never* arise by mere luck, and, of which Darwin knew nothing.
8. Life's *thousands* of specifically ordered protein sequences, which could never arise by chance, not just unlikely, it's impossible.
9. *Bird's Feathers and Avian Lungs,* neither of which could be created by random genetic (copying error) mutations (Denton).
10. The key *evidence of the fossils* which refutes Darwin.
11. The data of SUDDEN species emergence which refutes Darwin.
12. The tremendous *stability* of the species which refutes Darwin.
13. The evidence from *embryology* which contradicts Darwin.
14. The evidence from genetics and molecular biology which refutes Darwin's tree of life.[156] It's not a tree of life, as he envisaged it, but a wonderfully interconnected *web* of Life.
15. The evidence of the *world's most skillful breeder*s that it has never been possible to change any species beyond certain *genetic limits* without causing damage or death—never mind a new species.

Some thinkers are fond of quoting Occam's Razor: the principle that, in seeking to explain something, we should go for the most economical explanation. Yes, absolutely. So, when the most economical explanation for something is that, "If it *looks* purposeful, it probably *is* purposeful," or that, "If it *looks* intelligently designed, it probably *is* intelligently designed," perhaps we should take that possibility more seriously, than insisting on our *prior faith* in materialism? Or because we so firmly believe that we must never let a *spiritual* foot in the door?

14 Is Sudden Appearance Evolution?

"The ABSENCE OF FOSSIL EVIDENCE for intermediary stages between major transitions in organic design, indeed *our inability, even in our imagination,* to construct functional intermediates in many cases, has been a persistent … problem for gradualistic [Darwinian] accounts of evolution." So wrote world renowned fossil expert, S. J. Gould[157]

"DARWIN'S PREDICTION of rampant, albeit gradual, change affecting all lineages throughout time is REFUTED. The record ... speaks for tremendous *anatomical conservatism*. Change in the manner Darwin expected is *just not found* in the fossil record." World class fossil expert, Niles Eldredge.[158]

"[It] remains true, as every paleontologist knows, that most NEW SPECIES, genera, and families, and that nearly all new categories above the level of families, APPEAR in the [fossil] record SUDDENLY . . ." World class fossil authority, George Gaylord Simpson.[159]

These statements, by world class fossil experts, all *falsify* Darwin. Yet, their authors all remained essentially Darwinian in spirit. Why?

Because to their way of thinking, their statements didn't change very much. They still wanted to believe in Darwin's theory overall, and they continued to seek purely PHYSICAL theories to explain the data. Günter Bechly quotes from *The Comprehensive Guide to Science and Faith:*

> "Every theory makes certain predictions." Darwin's core prediction was *"gradualism,"* that all the changes in the history of life were due to *"a continuous accumulation* of small changes" over time. "Because Darwin wanted a wholly *naturalistic* explanation" for life's history, whereas, SUDDEN CHANGES in organisms "would require *miraculous* events.

> Therefore, he mentioned not fewer than six times in his magnum opus *On the Origin of Species* the Latin phrase *Natura non facit saltus." 'Nature does not make leaps.'*

Bechly, "This claim is still made by Darwinians today. The most well-known modern popularizer of Darwinism . . . Richard Dawkins, wrote in his 2009 bestselling book, *The Greatest Show on Earth,*

"Evolution not only is a *gradual process* as a matter of fact; it *has to be gradual* if it is to do any explanatory work. Gradualism is not an optional element of Darwinism, but it is essential for its success as a MATERIALIST or *anti-spiritualist* explanation for the origins and diversity of life. *If gradualism is wrong* Darwinism is falsified."[160]

Eldredge and Gould's fossil discoveries *falsified* Darwin. However, rather than accept this, they tried to rescue Darwin's theory from the fossil facts with an idea which they called punctuated equilibrium or PE.

PE tried to account for the fact that, by contrast with Darwin's claims, Life is *stable* for long periods, followed, at times, by *rapid change,* and by the mysteriously *sudden,* not gradual, appearances of new species. However, the problem with PE, is, that the relatively limited numbers—and harmful effects—of random genetic mutations could not possibly lead to the *vast changes* in form and function needed to make either PE or Darwin's more gradualistic ideas work.

Random genetic mutations, as Stephen Meyer explained in *Darwin's Doubt,* do not have the necessary creative power. Darwin said evolution needed "infinitely numerous transitional links" forming *"the finest graduated steps."* He said *"nature does not make leaps."*[161] The fossils contradict him. New species appear in an *abrupt, non-evolutionary,* way.

To those who say that the earlier Life forms gave rise to diverging families of species, there is *no evidence* for any such ancestors. Because as Colin Patterson explained to Tom Bethel, in the London Museum of Natural History, the *theoretical* spaces on the *theoretical* branches of Darwin's *theoretical* tree of evolution are *non-theoretically empty!*

There are imaginative drawings of Darwin's imagined transitional species, and the *faith* that his *theoretical* shared ancestors *'must'* have been there, because, *without* them, his theory fails. But this puts the cart before the horse, and, we can't ignore the fact of sudden appearance.

While the sudden appearances of the species is mysterious, consider the Big Bang: an E N T I R E U N I V E R S E appearing from, *seemingly,* no-where. It's not some slow, evolutionary thing So do some things in existence just *suddenly emerge?* This is the data we have, *sudden appearance* is the rule, no plausible ancestors, and no plausible transitional species. We need to remember too the strict limits to genetic change, which the most skilled breeders have never been able to cross. The data from embryology and genetics also does not support Darwin.

So what do we do? We work with the data we *have,* not the data we'd *like* to have, whether we favor Darwin, or whether we are young-earth creationists who try to impose Bishop Ussher's 17th century ideas onto the evidence. The evidence is that, in conflict with Ussher's ideas, Life on Earth is far older than the few thousand years he allowed for, based on his far too literal readings of the timelines in the bible.

The evidence is that, *in conflict with Darwin,* the species *suddenly emerged,* in a mysterious pattern of successive Life Waves, followed, at times, by mass extinctions. While this doesn't give us a new theory of Life, it *does* free us to consider the evidence as it *is,* and not try to force it to fit either Bishop Ussher or Darwin's ideas.

It is also useful to remember that in deep reality, as Max Planck the discoverer of quantum physics pointed out, there is **NO MATTER** as such. There is just e . n . e . r . g . y s . p . a . c . e 99.9..9..9..9..9..9..9..9..9..9..9..9..9% of it! And everything, *this* amazing second and *every* amazing second, is disappearing and reappearing, de-materialzing and re-materializing, out of the quantum ground, now here, now not here, *trillions* of times a second. Which is to say that, *all* things having FORM and imbued with *Living Being,* are emergences from the *Original, Form-less—It-Just-Is —Unseen,* to the **IN-FORM-ED,** to the *en-spirited* **MATERIAL,** and the *en-souled* **MECHANICAL.** Some in-**FORM**-ings, like the Big Bang, occur very rapidly, some, like the Cambrian Explosion, less rapidly, and, others take even or much slower forms.

The key question for us, though, is not in.FORM.ation, the data of which is all around us, from atom to GALAXY and everything in between, but are the *Plants, Animals, and People* intelligently in.FORM.ed phenomena, or are they just *Flower–Shaped, Animal–Shaped, and People–Shaped, flukes of chemistry,* with no more *Soul* or meaning to them than to the random rocks strewn around a dry valley?

Which makes more sense and which is more logical, overall? Do we live in a ***mysterious-but-dumb*** universe, as the materialist school urges, or in a ***mysterious-but-intelligent*** *one,* as all the great classical thinkers, from Buddha, Lao Tzu, Plato and Jesus, to Galileo, Newton, and Planck held? Are we part of an *intelligent cosmic mystery or* MINDLESS?

If the *amazing* materialization of an *entire universe,* from no-thing to some-thing, was possible, then other seemingly *non-evolutionary* emergences may also be possible. Again, our most basic question is simply this: do Life's amazing expressions come from an Intelligent Universe Generating Mechanism which ***'just is,'*** or from a MINDLESS Universe Generating Mechanism which ***'just is'****?*

Each of these is, of course, as *seemingly* impossible as the other, because, be the *Final Source of All,* ***intelligent,*** or, be it MINDLESS, *HOW can it exist? HOW does it exist? HOW can* there be a *cause* of anything at all, rather than nothing at all? Even one atom? No one has an answer to these deepest of mysteries—they are impossible to answer.

What we can ask, though, is: *which* of the two possible *Ultimate Causes* of things, *intelligent vs* MINDLESS, is, all things considered, the more likely? The answer seems obvious to me, and, I hope, to you too.

Our *atheistic* friends say there cannot *be* any such causes, because they are not *physically* visible: "We can't *see Spirits or Souls* in our telescopes!" On the other hand, our *spiritually* minded friends accept that Life's more *subtle* causes are not physically visible. But, *they* say, "We can *infer* them, using our reason, logic, intuition, and commonsense." They add, "Open your eyes! Look at a child's smile, or at your lover's eyes. Look at the eagle's wing, or think of the bee's sting. Consider the marvels of *Eye, Tongue, Blood, Liver, and Lung.* It is absurd to claim that any of these amazing things *could* arise, let alone *did* arise, by pure, *stupid,* chance." Why don't more of us see?

Meanwhile, regardless of anyone's pet theories, the fossils continue to disagree with Darwin, as they always have.

Firstly, because of, *"Sudden Appearance* – and secondly because of, *Stasis* – most species exhibit *no directional change* during their tenure on earth. They appear in the fossil record looking much the same as when they disappear; morphological *change is usually limited and directionless."*[162] Which is another way to say that there is NO EVIDENCE FOR Darwinian EVOLUTION. Yet, those words were written, not by me, but by world famous fossil expert S. J. Gould.

In 1859, Darwin published, what he said, were just some "BRIEF EXTRACTS" from much larger manuscripts, "imperfect," *without* references, only "general conclusions," with just a "FEW FACTS in illustration." These "extracts," with just a "few facts in illustration," can no longer be considered good science. A "FEW FACTS" were not nearly enough to justify the towering edifice of speculation upon speculation, *'might* this, *might* that, over and over again,' that he and his supporters erected after he published—this now tottering, rotting structure. Especially, as his promised "future work" to back them up, "in detail," with "ALL THE FACTS," and references,[163] never came.

S. J. Gould, a world class authority, pointed out, all the way back in 1980, that the modern version of Darwinism, which is called neo-Darwinism—supposedly driven by random genetic copying error mutations—was already *"dead,* despite its persistence as textbook orthodoxy."[164] Dr Antony Latham agrees. "The general situation found in successive [fossil] strata is that of *stasis,"* not Darwinian evolution.

On the back-boneless animal body plan, called echinoderms, to which Star Fish and Sea Urchins belong, Latham comments,

> "There are NO TRANSITIONAL FORMS found in the fossil record that link the various types of echinoderm since

> their first appearance in the Cambrian period. What we see is the initial phylum [anatomical design] APPEARING SUDDENLY around 530 million years ago – in a wide variety of forms . . . I have searched hard for any evidence of transitional forms between the major groups which exist today and I have SEEN NONE. What we do see is [just] small changes in existing groups (micro-evolution).
>
> 'The initial phylum appears along with all the others in the Cambrian explosion, it is fully complex and has more classes than we have now; the classes that survive appear suddenly also. The only clear evolution is on the micro-scale. It is the sudden appearance of new forms without linking transitionals that remains the real mystery – just as it was in Darwin's day.
>
> 'All that is written about the echinoderms is mirrored in the fossil record of the other invertebrates. I have used the echinoderms as one example of what is found also amongst fossils of the other main phyla of invertebrates: arthropods, mollusks, sponges, cnidarians, bryozoans and brachiopods.
>
> 'Check any textbook on the subject and you will see that SUDDEN APPEARANCE and DISCONTINUITY is the norm. The problem that Darwin so clearly saw has not gone away with time and the FOSSIL RECORD seems now to REFUTE THE ENTIRE BASIS of his theory of gradual changes." [165]

Earlier, in 1979, Luther Sunderland wrote to the then senior fossil expert at the Natural History Museum in London, Colin Patterson, to ask him why, in his then recent book, *Evolution,*[166] he had *not* included a single photo of a *transitional fossil* demonstrating Darwinian evolution. Patterson replied with this rare, but honest, admission from a professional evolution academic. He wrote,

> 'I fully agree with your comments on the lack of direct illustration of evolutionary transitions in my book. If I *knew* of any, fossil or living, I would certainly have included them. You

suggest that an artist should be used to visualize such transformations, but where would he get the information from?

'I could not, honestly, provide it, and if I were to leave it to artistic license, would that not *mislead* the reader?' Patterson added:

> '[S. J.] Gould and the American Museum people are hard to contradict when they say there are NO TRANSITIONAL FOSSILS. ... You say that I should at least "show a photo of the fossil from which each type of organism was derived." I will lay it on the line—there is NOT ONE SUCH FOSSIL for which one could make a watertight argument.'[167]

'"Not one such fossil. . ." Here was yet another world-class fossil expert admitting that the data didn't *actually* support Darwin's famous theory. Patterson was following the data—which is what Eldredge and Gould were trying to do with their theory of punctuated equilibrium PE.

So what was PE? It was their idea that the species will be *stable* for long periods, *without* changing in the way Darwin predicted, but, when evolutionary change *did* occur, *it was so rare* that *no fossil evidence* will ever be found for it. Their solution to *the failure* of Darwin's theory may sound fanciful, but Eldredge and Gould were not anti-Darwinian. They were just trying to *keep* their Darwinian cake—and to eat it.

They admitted that there are "NO LINKING FOSSILS," which, in any less contested area of science, would have refuted Darwin's entire thesis, and allowed us all to move on. They were, however, reluctant to follow the logic of their own discoveries and to let Darwin go.

So why did they continue to argue that Darwin was *'right,'* despite their own findings that he was *wrong?* Shouldn't they have started over?

Yes, but they claimed that when the species *did* transform—*ra*N*d0*m microbe-to-*ra*N*do*m-worm, random-microbe-to-*ra*N*d0*m-snail, random-microbe-to-*ra*N*do*m-fish, random-fish-to-*ra*N*do*m-amphibian, *ra*N*d0*m-dinosaur-to-random-bird, random-bear-to-random-whale, random-shrew-to-random-bat, random-ape-to-random-man—there were *so few* transitional forms that *no fossil examples can ever be found.* Darwinian evolution *did* happen, *honestly!,* it did, but it happened in such huge JUMPS that there were *just too few* transitional species for there ever to be any preserved *transitional* fossils showing Darwin's process in action. But isn't science supposed to *follow* the data?, not ignore it?

Also, problematic for Eldredge and Gould, is that they had no plausible *mechanism* to drive the very *rapid* changes they argued for, *versus* Darwin's slow—nature does *not make leaps*—form of evolution.

Darwin and his followers have repeatedly tried to force the data to fit with his theory, but, in truth, they have lost the argument.

Contemporary, thinkers like Richard Dawkins and Richard Lewontin claim to be the defenders of *'science,'* but all they are really defending is the modern MATERIALIST (atheist) belief system with its strange, *"pure-luck-can-create-all"* dogmas, urging us to leave our brains at the door, and to ignore that *every, single, cell* in our bodies is vastly cleverer than anything we can make, using all our powers.

Darwin's, *"it's-all-pure-luck,"* scheme has not explained *Life.* It is now damaging science, not furthering it. This is why the sooner we free ourselves from his paradigm of unintelligent design, the better.

As science writer Richard Milton notes in ***Shattering the Myths of Darwinism: a Rational Criticism of Evolution,***

> "Darwin's theory of evolution relies on the idea of random genetic mutations… altering species... But… the extent of genetic change wrought by natural selection is quite limited…
>
> "[It is also an] embarrassing fact that… NO TRANSITIONAL SPECIES SHOWING EVOLUTION IN PROGRESS BETWEEN TWO STAGES HAS EVER BEEN FOUND… the theory of evolution has become an act of *faith* rather than a functioning science…
>
> "Not until the scientific method is [properly] applied to Darwinism will it be exposed, and only then will the right questions be asked about the mystery of life on earth."[168]

15 Fabulous Bats

'In *Evolution, a Theory in Crisis,* biologist Michael Denton writes,

> "The difficulty of envisaging how evolutionary gaps were closed [by Darwinian processes] does not stop with birds: take the case of the bats. The *first known bat* which [suddenly] appeared in the fossil record some 60 million years ago had as completely developed wings as modern forms. As in the case of birds, how could the development of the bats' wings and capacity for powered flight have come about gradually?" [169]

Bats, like the other species, *appeared abruptly* on our planet.

There no fossils of un-bats on the way to becoming bat-like.

No fossils of wingless rodents gradually turning into *fast-flying, sonar-equipped* bats. In *Billions of Missing Links: A Rational Look at the Mysteries Evolution Can't Explain,* G.S. Simmons writes,

> 'Biologists speculate that the bat's ancestors might have been tree dwelling shrews or moles who had some capacity to glide, but that's mere conjecture [a Darwinian *"Just So"* story]. It doesn't *explain echolocation,* which is their [amazing] sonar ability to find food on the wing at night, or how they can purposefully drop their body temperatures at night to save energy. Most mammals constantly burn energy to maintain a specific temperature, 24/7.
>
> '... compared, by size, to our sonar detectors, theirs is one trillion times more efficient. Bats can fly at speeds up to 60 mph and at heights of 10,000 feet. Often they live in caves by the millions. Nurseries may have as many as 2000 pups per square meter, yet mom always seems to know her offspring. One interesting capability that breaks the rules of [Darwinian]

evolution is the bat's ability to *lock onto a rock* and hang upside down without falling. This is made possible by an automatic locking mechanism in their feet. A tendon closes the toes against the rock, and then the bat's weight creates the proper tension downward to keep it locked. The moment the bat grabs the rock the clamping mechanism snaps in place.

> 'The *unprecedented* backward facing knees with forward facing feet aid the process. Bats will sleep and carry on most of their necessary activities in a hanging position for hours, only flipping 180° whenever there's a need to evacuate their bladder or defecate. One might *feel sorry* for the predecessors who had not yet learned to flip over.
>
> 'When it's time to hunt for food they just they just release, drop, and take off. Partially-hanging, sometimes-falling intermediates did not exist.'[170] [There are no transitional bat species.]

'Darwin said that, "if his theory be true", the fossil records ought to show existing species G R A D U A L L Y turning into new species,

> "by differences not greater than we see between the natural and domestic varieties of the same species at the present day. [... and] the number of intermediate and transitional links, between all living and extinct species, must have been inconceivably great. But assuredly, *if this theory be true,* such have lived upon the earth."[171]

"If this theory be true..." Ah, we do so hate to admit that *we don't know* and that, despite all the claims, Darwin's, once promising (to some) theory has failed. Darwin wrote,

> "The number of intermediate varieties, which have formerly existed on the earth, must be truly *enormous.* Why then is not *every* geological formation and <u>*every*</u> stratum <u>*full*</u> of such intermediate links? ['<u>*Every?*</u>' '<u>*Full?*</u>' There are *none!*] Geology assuredly does not reveal any such finely graduated organic

> chain;" [hold on, it reveals no such chain *at all!*] "and this, perhaps, is the most obvious and GRAVEST OBJECTION which can be urged against my theory."[172] [Finally, some truth and reality.]

Because, far from providing a "finely graduated chain" and an "inconceivably great" number of fossil links "between all living and extinct species," the fossils provide *no such links at all.*

Darwin's technique with objections was to make excuses as to why, despite the fossil data *not* agreeing with him, his theory *'ought'* to be right because he felt it was such a good one. Yet, at the same time, he said, we need to reject his theory, unless we *truly* think that the *lack* of convincing *linking* fossils, the *sudden,* not gradual, species appearances and the *limited* powers of natural selection all don't matter.

However, rather than following the *actual* fossil data, Darwin, and his fans, have always tried to force the data to fit in with his ideas, rather than going back to basics and asking the right questions.

Could life arise by chance? Could the random mixing of NON-LIVING CHEMICALS *really* create *Cells and Shells, Wings and Feathers!, Livers and Kidneys, Hearts! and Lungs!, Muscles and Sinews, Hands and Fingers, Eyes and Brains—for no reason at all?*

Is blind LUCK *really a supreme creative principle?* Would chemistry, genetics, and, above all, the *fossil evidence,* eventually, confirm his theory? Did Life on Earth unfold as he predicted? Did non-flying rodents really "evolve," by pure chance, into very different Bats?

Luther Sunderland says, in *Darwin's Enigma,* that the fossils really frustrated Darwin, "because they *did not reveal any evidence* of a gradual… evolution of life from a common ancestor, *proof which he needed* to support his theory."[173] Darwin just hoped that "future explorations would… prove his theory correct."

In the meantime, he tried to explain away the *unhelpful* fossil records by claiming, "The geological record is *extremely imperfect* and this fact will to a large extent explain why *we do not find* intermediate varieties, connecting together all the extinct and existing forms of life by the finest graduated steps ['finest graduated steps?' Dear Darwin, we do not find any such fossils at all]." Darwin continued,

"He who rejects these views on the [poor] nature of the geological record, WILL RIGHTLY REJECT MY WHOLE THEORY."[174]

So why haven't we "*rightly rejected* Darwin's entire theory?"

Why do we *take no notice* of the many statements similar to this:

"Darwin's prediction of rampant, albeit gradual, change... is REFUTED. ... Change in the manner Darwin expected is *just not found* in the fossil record." Eldredge.[175] And this, "[It] remains true, as every paleontologist knows, that most NEW SPECIES, genera, and families, and nearly all new categories above the level of families, APPEAR in the [fossil] record SUDDENLY . . ." Simpson.[176]

And this, *"There are no transitional fossils."* Patterson.[177]

And this, *"No transitional species* has ever been found... the theory of *evolution has become an act of faith...*" Milton.[178] Why do we take no notice? Because, for so many of us today, Darwin's theory has become its *own kind of 'religion,'* which cannot be questioned, without unjustified accusations of being *unscientific,* or a biblical literalist.

Lyn Margulis, a distinguished biologist, agreed that Darwin had not successfully explained natural history. She wrote, "Natural selection eliminates and maybe maintains, *but it doesn't create."*[179]

> 'One [random genetic] mutation confers resistance to malaria but [it] also makes happy blood cells into the deficient oxygen carriers of *sickle cell* anemics. Another converts a gorgeous newborn into a *cystic fibrosis* patient or a victim of early onset *diabetes.*
>
> 'One mutation causes a flighty red-eyed fruit fly to fail to take wing. Never, however, did that one mutation make a wing, fruit, woody stem, or claw appear. Mutations, in summary, tend to induce sickness, death or deficiencies.
>
> *'No evidence* in the vast literature of heredity changes shows unambiguous evidence that random mutation itself, [RGMs are genetic typos,] even with geographical isolation of

populations, leads to speciation. Then *how* do new species come into being?'[180]

In an interview she said,

> "This is the issue I have with neo-Darwinists: They teach that what is generating novelty is the accumulation of random [genetic] mutations in DNA, in a direction set by natural selection. If you want bigger eggs, you keep selecting the hens that are laying the biggest eggs, and you get bigger and bigger eggs. But you also get hens with defective feathers and wobbly legs. Natural selection eliminates and maybe maintains, but it DOESN'T CREATE … neo-Darwinists say that new species emerge when mutations [genetic typos] occur and modify an organism. I was taught over and over again that the accumulation of random mutations led to evolutionary change – led to new species. I BELIEVED IT, until I looked for evidence." [181]

Doubts about Darwin's ideas are not new. At times, he had his own grave doubts. He just hoped that, *in the end,* he'd be proved right.

But his claims for *natural selection,* to create and to evolve all Life, were what he thought *might* be true, not the evidence he needed.

Hence, he *bluffed and bluffed,* promising he would prove his claims in a later, greater work, which, it is no surprise, he never published.

One of Darwin's followers, Ernst Mayr, wrote,

"It must be admitted, however, that it is *a considerable strain on one's credulity* to assume that such finely balanced systems such as certain sense organs (the eye of vertebrates, or the bird's feather [or lungs, or livers, or the senses of smell and hearing, or blood, or hearts or brains]) could be improved by random mutations."[182] Ah yes! How true!

Malcolm Muggeridge, a well known 20th century intellectual, was less charitable. He observed that Darwin's "theory of evolution," which, we are told is *'science gospel,'* and not to be questioned, "will be one of the *greatest jokes* in the history books of the future. Posterity will marvel that so very flimsy and dubious an hypothesis could be accepted with the incredible *credulity* it has [and for so long as it has]." [183]

Darwin wanted something more scientific than: "Just read the Bible, Creation is described there…" To this extent he was right. the scriptures are not *'science.'* They are symbolizations of the classical, logical, inference that *Life and Mind, our Love for Goodness, Beauty and Truth, our Intelligence, our Curiosity, and our Creativity's ultimate sources are meaningful and intelligent,* not MINDLESS or RANDOM—that's all.

Unfortunately, Darwin and his friends, having correctly rejected the creation story in the bible, as 'not-science,' (of course it wasn't!) went to the *other* extreme of trying to find totally CHANCE-BASED, *non-intelligent* explanations for all *Living things,* including, even, for our very *Minds, and Souls, and feelings*—which *if* it worked, would be fine.

But Darwin, and many since he, have conflated the rightness, or wrongness, of traditional religious views with the more basic question as to whether the *supernatural* exists at all. Yet, ironically, the *spiritual* and the *supernatural* are real—provenly so.[184]

However, while the religions have the *basic fact* of the *supernatural* right, they may well be wrong on many of the details. Their unrealistic ideas of eternal hells and boringly pious heavens, for example. All too often they portrayed *Spirit* as this grim, punitive force. Is it any wonder, then, that so many of us have left their teachings?

They ask us to believe, yet refuse to take any notice of the modern *spiritualist* research which challenges some of their ideas. But their limitations should not stop us from realizing that there is far *more* to Life than our materialist and atheistic friends are willing to allow.

There is, clearly, *a vast intelligence* to the *moment!, by moment!, the second!, by second!,* arising of *Living Things, breathing!, feeling!, sensing!, hearts beating!, blood circulating! kidneys treating!,* trillions of biochemical events taking place in our bodies, *right now!, and now!, and now!, this second!, and every amazing second!*

We need to *follow* the data, not to keep fighting it, as too many of Darwin's fans have urged us to do. Of course, prior to Darwin, it was accepted that many things, especially *Living things,* had facets beyond their purely physical components, like the *Life Forces* and the *Soul.*

It was what most of the world's greatest thinkers had always held. It was only during the 19th century that the dismal spirit of *materialism*—what the great Austrian philosopher, Rudolf Steiner, referred to as *Ahriman*—gained such a grim sway over the western academic mind.

So many now do not understand that just because we have some understandable frustrations with the traditional religions, this does not mean that the *Supernatural* and the *Spiritual* do not exist at all, as thinkers like Darwin, Dawkins, and Lewontin have urged us to believe —as though the limitations of the religions prove *materialism* to be true.

Biologist, Richard Lewontin, argues that scientists should be willing to "to accept scientific claims that are against common sense," and to tolerate "unsubstantiated just so stories," like Darwin's scheme, because, he says, science is in a struggle with the supernatural," and that scientists should have "a prior commitment to materialism."[185]

It is odd, though, that it doesn't occur to him that the *physical nature* and the *laws of physical nature*, which he so respects, may not stem from a *mysterious Cosmic Intelligence which '**just Is**,* no less than the mindless MATTER, and its mindless LAWS, which he believes in, ***just is***.

Nor does he allow that, far from being violated by this *subtle, all-pervading—atoms-to-stars—Cosmic Intelligence*—which ***just Is***—the amazing laws of nature are *generated by* this *same intelligence,* to which some of us give religious names, the *logical* existence of which he is so keen to dismiss, such is his *faith, his superstition,* in the powers of blind chance, which ***just is***, to create all, for no reason at all.

Also, while commonsense is not always the best guide, to rule it out of our scientific thinking altogether would be odd, wouldn't it?

The classical, common sense hypothesis—which thinkers like Darwin, Crick, Dawkins, Dennett and Lewontin all reject—is that neither Life nor Mind could *possibly* arise by chance, they are far too complex for that. These well know thinkers rejection of the *supernatural* is not intellectually, or logically, or scientifically justified.

16 Icons of Evolution or Fakes of Evolution?

We are told that Darwinian evolution is a fact. However, in *Icons of Evolution: Science or Myth?,* Jonathan Wells examines the surprisingly few 'facts' Darwinians have had to hand to prop up their shaky theory, and he shows how they are all, either misleading, or false, or fake.

For example, Miller and Urey's famous 1950s origins of life experiments, which we were led to believe were proof that Life might arise by chance, when all the experiments produced were a few, random, amino acids. Well, it's true that amino acids are *among* the basic building blocks of life, but building blocks, unguided by higher level forces will, it should be utterly obvious, never build anything functional.

In *Evolution's Final Days,* John Morrison asks, "If we threw a pile of blocks on the ground would they land in an ordered pile? Of course not. We could throw the same blocks for ever. We would never get an ordered pile." Of course not! As Morrison, rightly, asks:

"Why are the laws of probability taken into account in all other areas of science, but *not* when it comes to evolution? Darwin's theory of evolution states that random amino acids somehow came together, [by pure, unintelligent, unguided, luck,] thus violating the 2nd law of thermodynamics [entropy] and became a [complex] cell.

"Then, this single-celled, asexual organism randomly mutated and... has since *become [ever] more ordered* with every mutation." Darwin's fans appeal to vast time to justify this impossible idea, but as Denton says, *"time in itself tells us nothing* of the probability of achieving any sort of goal unless the complexity of the search can be *quantified."* [186] Yet Darwin did not even try. *Why? Why* did he not even try?'

The next icon Wells considers is Darwin's famous tree of evolution. If biology textbooks were to be more candid, and more accurate, they would explain that Darwin's fantastical, unrealistic [impossible is better] notion of tiny microbe, to vast tree of Life remains unproven. In fact, some Darwinians are starting to say that the 'evolution' they now believe in is not tree-like at all, but web- or network-like.[187] But that's not Darwinian evolution. Because, in a network or web-of-life, there is no obvious starting place or progression, is there?

Another Darwinian icon relates to homology, meaning the interesting similarities in skeletal structures shared by many vertebrates. For example, the similar forelimbs of bats, porpoises, horses and people, including the common five finger pattern.

Prior to Darwin, these similarities were seen as design archetypes of the *same, spiritual, intelligence* which gives rise to everything. Darwin, by contrast, asked: "What if *Living* things were *not* generated by any kind of *intelligence?* Perhaps, microbial life might arise, by chance, in a 'little pond,' and everything else come from that?" It was a radical idea.

But was it correct? Well… (i) it did *not* actually show how *Life* started. (ii) Not *even a few,* let alone the vast numbers of linking fossils his theory called for are ever to be found.[188] (iii) There is no believable *non-intelligent* mechanism which could drive the *highly progressive evolutionary* processes Darwin had in mind.

Yet, Wells says, the standard text books on Darwinian Evolution omit these key points. Nor do they explain that both of the mechanisms proposed to account for these homologies—similar genes and similar developmental pathways after conception—*fail.* They do not point towards common descent but to *separate* lines of descent, which points, more logically, to intelligent causation, not to rAnDo*M*.

Darwin did not know of genes, but his modern supporters, who do, claim that these similarities arise, either, (a) due to *similar genes* inherited from a common ancestor, or, (b) due to *similar growth pathways* following conception. *If* it can be *shown,* then, that similar structures in different species are produced by similar genes, this would be supportive of Darwin's key idea of a branching, tree-like descent from a universally shared ancestor. "But," Wells writes, *"this is not the case* and biologists have known it for decades. In fact, all the way back in 1971 Gavin de Beer wrote: "Because homology [similarity of form] implies *community of descent* from . . . a [shared or] common ancestor it might be thought that genetics would provide the key to the problem of homology [skeletal similarities]. This is where the *worst shock of all* is encountered . . . [because] characters controlled by identical genes are not necessarily homologous [similar] ...homologous structures need not be controlled by identical genes."[189]

Equally striking, Wells writes, is the fact that not only do *different* genes give rise to *similar* structures but *similar* genes can give rise to *different* structures. "Geneticists have found that many of the genes required for proper development in fruit flies are similar to genes in mice, sea urchins and even worms. In fact, gene transplant experiments have shown that developmental genes from mice (and humans) can functionally replace their counterparts in flies." [190]

Darwin's supporters argue that similar biological structures must have arisen due to a *shared* evolutionary history. If true, this ought to be reflected in *similar* growth pathways following conception. However, as so often with Darwinism, their ideas do not match the real world data.

Yet, "It is a familiar fact," wrote embryologist Edmund Wilson, in 1894, "that parts which closely agree in the adult, and are undoubtedly homologous [similar], often *differ widely* in larval or in embryonic origin either in mode of formation or in position or in both."

Wells comments, "Gavin de Beer agreed: "The fact is that correspondence between homologous [similar] structures [in the adult forms] *cannot be* pressed back to similarity of position of the cells in the embryo, or of the parts of the egg out of which the structures are ultimately composed, or of developmental mechanisms by which they are formed."[191]

Yet, Wells notes, most biology textbooks *fail to point out* that the two mechanisms proposed to account for the similarities in vertebrate limbs *do not* support Darwin. So, he asks, if the continuity of information required to build up vertebrate limbs comes *neither* from genes, *nor* developmental pathways following conception, *how* can it come from a universally shared descent? Moreover, we *haven't a clue* how this incredibly complex genetic information arises in the first place. Although the data argues for *separate* lines of descent, *not* shared lines of descent, this is resisted because it conflicts with Darwin's ideas. But, dear reader, have you noticed a pattern here? If you answered, "Yes, it's always *'the evidence that's wrong,'* never Darwin," *10/10!*

Unfortunately, for too long, Darwin's ideas have been treated like *'revelations'* providing *'answers'* to Life's mysteries which cannot be questioned. This is *not* how science is supposed to be. Science can also be about *accepting that there are questions* to which do *not* yet have answers, even as we remain curious and continue to explore.

Science is not pretending we *'know'* when we don't.

Another dodgy Darwinian icon is Ernst Haeckel's idea that *'ontogeny* recapitulates *phylogeny,* which is the idea that the movement of water breathing creatures, out of the sea, onto the land, via an *amphibious,* water-breathing to air-breathing stage, may, even now, be seen to be recapitulated in the early stages of embryos, including what, at first glance, can appear to look like *vestigial gill slits*. In other words demonstrating a fish ancestry.

To back up his idea, Haeckel produced a set of drawings of different embryos, including fish, salamander, tortoise, chick, rabbit and human, seeking to show how similar they all are, especially early on. Sadly, to his discredit as a scientist, Haeckel *doctored* his drawings to make the embryos *look more alike* than they really are. He did this to bolster Darwin's idea of shared descent. In other words it was a *bluff-scam.*

Regrettably, what Darwin believed to be the supporting evidence from embryology, as misrepresented to him by *Haeckel,* gave him more confidence in his theory. He didn't know that Haeckel had doctored his drawings, and Darwin said he found them very persuasive. He wrote, "it is probable, from what we know of the embryos of mammals, birds, fishes and reptiles, that these animals are the modified descendants of some ancient progenitor they had in common." Wells explains that since Haeckel's embryo drawings *seem* to "provide such powerful evidence for Darwin's theory . . . some version of them can be found in almost every modern textbook dealing with evolution." Yet biologists have known for a long time now that Haeckel *faked* his drawings. Embryos *never* look as similar as he claimed.

"Although you might never know it, Darwin's "strongest single class of facts" is a classic example of how evidence can be twisted to fit a theory.'[192] Wells comments that if one started with the evidence, then examined Darwin's ideas, one would conclude that the various classes of vertebrates, *rather than* being descended from a common ancestor, probably had *separate origins.*

He notes that Von Baer, a famous 19th century embryologist, couldn't support Darwin *precisely because* Darwin's ideas conflicted with the real-world data from embryology. Wells continues,

> "Many modern Darwinists haven't changed. It doesn't matter how much the *embryological evidence* conflicts with [their] evolutionary theory – the theory, it seems, must not be questioned. This is why, despite *repeated disconfirmation,* Haeckel's biogenetic law and [his] faked drawings haven't gone away.'[193]

S. J. Gould called Haeckel's faked embryo drawings "the academic equivalent of murder," a serious charge. Yet, they are still around, so, many people, including other scientists, who are not experts in this field, continue to be misled.

Wells argues, and I hope you agree, that more accurate textbooks would *not* recycle misleading drawings of embryos, would *not* call pharyngeal pouches 'gill slits,' as though these pouches are evidence of fish ancestry (they have nothing to do with fish gills!), and would explain to students of natural history that *embryology tends to refute,* not to confirm, Darwin's ideas.

Yet, Wells says, many textbooks continue to fail to explain that earlier embryonic stages are *dissimilar,* and they perpetuate the misleading claim that embryology provides evidence for Darwin's idea of common ancestry.

The next dodgy icon Wells discusses is *Archaeopteryx.* Rather than questioning the, by now, dubious and unrealistic claim that this ancient bird provides evidence for a transitional link between reptiles and birds, many textbooks give no clue that there is any controversy over its ancestry, nor do they point out that modern birds are almost certainly *not* descended from it.

Wells also looks at Darwinism's famous, minor, backwards and forwards, finch beak variations, where, rather than explaining that selection on finch beaks oscillates between wet years and dry years *without* resulting in any net evolutionary gain, many books report these minor, backwards and forwards (reversible) variations as if they are evidence for Darwin's hypothesis of long-range, macro-evolution.

Wells goes into other Darwinian icons such as peppered moths, four winged fruit flies, the evidence from horse fossils, and theories about *human origins.* The last of which are "controversial, and rest on little evidence." All the monkey-to-man drawings of ape-like human ancestors are hypothetical, imaginative, yes, but of no scientific validity.

Casey Luskin adds that, while some claim that the evidence for humans having evolved from ape-like ancestors is overwhelming, in fact,

> 'hominin fossils generally fall into one of two categories — ape-like species or human-like species (of the genus *Homo*) – and there is a large UNBRIDGED GAP between them. ... the fragmented hominin fossil record *does not* document the evolution of humans from ape-like precursors. In fact, scientists are quite sharply divided over who or what our human ancestors even were.
>
> 'Newly discovered fossils are often initially presented to the public with great enthusiasm and fanfare, but once cooler heads prevail, their status as human evolutionary ancestors is invariably called into question.
>
> 'The details of the earliest stages of human origins are murky. They come from what UC–Berkeley paleo-

anthropologist Tim White once called *"a black hole* in the fossil record."[194] ... While early hominin fossils are controversial, due to their fragmented condition, there is one major group – the *australopithecines* – that is *widely promoted* as directly ancestral to humans. The primary claim is that australopithecines had the head of a chimpanzee but a body that allowed it to walk upright, like humans.

'Despite the prevalence of that standard view, authorities have found that the fingers, arms, chest, hand bones, striding gait, shoulders, abdomen, inner-ear canals, developmental patterns, toes, and teeth of australopithecines *point away from their being human ancestors* and / or suggest that they didn't have human-like bipedal locomotion.[195] [This evidence includes the world famous australopithecine referred to as 'Lucy.'] ...

'Paleoanthropologist Leslie Aiello, who served as head of the anthropology department at University College London, stated that when it comes to locomotion, *"Australopithecines are like apes, and the Homo group are like humans.* Something major occurred when *Homo* evolved, and it wasn't just in the brain."'[196]

On the Big Bang-like appearance of humanity,

'When the human-like members of our genus *Homo* appear [in the fossil records], they do so *abruptly.* A paper in the Journal of Molecular Biology and Evolution called the appearance of *Homo sapiens* a *"genetic revolution"* in which "no australo-pithecine species is obviously transitional."[197] In a 2004 book, the famed evolutionary biologist Ernst Mayr explained that "the earliest fossils of *Homo*, *Homo rudolfensis* and *Homo erectus* are separated from *Australopithecus* by a large, unbridged G A P" with no "fossils that can serve as missing links."[198]

> '... the fossil record provides us with ape-like australopithecines ("before") and human-like *Homo* ("after"), but *not* with fossils documenting a transition between them. In the absence of intermediaries, we're left with "inferences" of a transition based strictly upon the [PRIOR] ASSUMPTION of Darwinian evolution.
>
> 'No wonder one commentator argued that IF WE TAKE THE FOSSIL RECORD AT FACE VALUE, it implies a *"big bang theory"* of the appearance of our genus *Homo*. [199] ... the evidence shows that human-like forms appear abruptly in the fossil record, without any fossils connecting us to our alleged ape-like evolutionary ancestors.
>
> 'This contradicts the expectations of neo-Darwinian evolution [common descent] and suggests that unguided [non-intelligent, non-spiritual, blind-chance] evolutionary mechanisms do not account for the origin of our species.'[200]

The comment: "If we take the fossil record at face value, it *implies a big bang theory* of the appearance of humanity," is interesting. Because, why would we *not* take the evidence at face value? Unless, we have decided that a prior theory, like Darwin's is already true, when that is the very thing in question: is it true?

As so often happens with *Darwin* in particular, and with materialism in general, too many prior assumptions go unquestioned. Sadly, too, Darwin, and too many of his supporters, take the line that his ideas are *'too good'* not to be true, which is not science but *ideology.* This is the issue with Darwinism and materialism today. They are no longer treated as scientific hypotheses open to being tested and, potentially, falsified but, rather, as settled *'scientific'* truths which cannot be questioned.

There is, though, far more that separates people from apes than just differences in gait, shoulders, chest shape, ear canals and teeth. Take, for example, our capacities for abstract thought, language and speech. In his book, *The Kingdom of Speech,* Tom Wolfe observes that not only is the human capacity to speak unique, but that, despite a vast amount of work, over decades, "by some of the greatest minds in academia," all evolutionary theories about the origins of human language are incredible, and they fail to explain these most unique features of what it is to be human. Michael Egnor writes,

"... Darwinists are at an utter loss to explain how language... "evolved." ... [No] evolutionary *just-so stories* come anywhere close to explaining how man might have acquired the *astonishing* [but, *biologically, totally unnecessary*] ability to craft unlimited [abstract] propositions and concepts and subtleties within subtleties using a system of grammar and abstract designators (words) …

"Darwin and his progeny have had no dearth of fanciful guesses – birdsongs (Darwin's favorite theory) and grunts and grimaces that mutate... into Cicero and Shakespeare. Evolutionary theorizing about language has been a colossal waste of time.

"None of this evolutionary fancifulness makes any sense, nor has any real scientific basis… The human mind has *immaterial* abilities – the intellect's ability to *grasp abstract universal concepts* divorced from any particular [concrete] thing—and that this ability makes us more different from apes than apes are from viruses. …

"We are not just animals who talk. … Language is the tool by which we think abstractly. ... (1) Language is a rock against which [Darwinian] evolutionary theory wrecks, one of the many rocks – (2) the uncooperative fossil record, (3) the jumbled molecular evolutionary tree, (4) irreducible complexity, (5) intricate intracellular design, (6) the genetic code, (7) the collapsing myth of junk DNA, (8) the *immaterial human mind* – that comprise the shoal that is sinking Darwin's Victorian fable." [201]

Towards the end of *Icons of Evolution,* in a chapter headed *Science or Myth?,* Wells quotes Ernst Mayr, who having once written that, "it is a *considerable strain* on one's *credulity* to assume that . . . systems such as … the eye of vertebrates, or the bird's feather…could be improved [let alone created] by random mutations,"[202] proposed, in *Scientific American,* that no educated person doubts any longer the "so-called theory of evolution, which we now know to be *a simple fact.*" Wells responds, He writes: "Ask any educated person *how* we know that evolution is *a simple fact* . . . and chances are that person will list some or all of the [questionable] icons described in this book. [Yet] For most people—including most biologists—the alleged icons *are* the very *'evidence'* for Darwinian evolution which they rely on—just as Darwin relied on Haeckel's faked embryo drawings."

However, these *'icons'* of evolution *misrepresent* the evidence. "One icon (Miller–Urey) gives the false impression that scientists have demonstrated an important first step in the origin of life.

"One (the four–winged fruit fly [a deformed and functionally useless byproduct of intelligently designed lab experiments]) is portrayed as though it were raw materials for evolution, but is . . . an evolutionary *dead end.* Three icons (vertebrate limbs, *Archaeopteryx,* and Darwin's finches) show actual evidence but are typically *used to conceal* fundamental problems in its interpretation.

"Three (the tree of life, fossil horses and human origins) are incarnations of [materialist] concepts *masquerading* as neutral descriptions of nature. And two icons (Haeckel's embryos, and peppered moths on tree trunks) are *fakes.*

"People such as Ernst Mayr insist that there is overwhelming evidence for Darwin's theory. . . . [Yet,] If there is such overwhelming evidence for Darwinian evolution, why do our biology textbooks, science magazines and television nature documentaries keep recycling the *same tired old myths?*

"There is a pattern here, and it demands an explanation. Instead of continually testing their theory against the evidence, as scientists are supposed to do, some Darwinists *consistently ignore,* explain away, or misrepresent the biological facts in order to promote their theory."[203]

Unfortunately, with Darwin, and too many of his supporters, "It is *always the data that's 'wrong'* never the theory. The fossils are wrong. The cell, or blood, or eye, *'might'* have evolved by chance, *mightn't it?*

"Please suspend your reason and your logical *disbelief…* just try to *imagine that Cells!, or Blood!, or Hearts!, or Brains!, 'might'* have evolved in tiny steps, by *pure chance!"* Well, we can *imagine* whatever we like, but Darwin, and his advocates, cannot seem to tell the difference between a *hypothesis* and whether, or not, it is *supported* by the data. Because of this, Darwin's theory continues to be treated as if it is *far* better than it is. But that's not science, it's *belief,* and it is not how science is supposed to be. Wells agrees. He writes,

"One isolated example of such behavior might be due simply to overzealousness. Maybe even two. But ten? Year after year?""[204]

In his review of *Icons of Evolution,* Dean Kenyon, professor of biology, reflects that Wells has "exposed the *exaggerated claims* and *[self?]* deceptions that have persisted in standard textbook discussions of biological origins for many decades, in spite of [all the] contrary evidence!" As Milton, Denton, Johnson, Wells, Latham, Behe, Dembski, Meyer, Bethell, Shedinger, and so many others, have shown, it is long past time to let Darwin go. Darwin hoped he might have discovered a new natural law. The branching, tree-like descent, with VAST changes, of all plants and animals, from just one or two tiny, microbial ancestors like bacteria: tiny-tiny-tiny-bacterium-to-worm-to-fish-to-dinosaur-to-bear-to-whale. *Honestly? By mindless chance?*

HOW does this make sense, when bacteria were, and remain, *one of the most, possibly THE most, successful Life Form,* with no rational biochemical or genetic need, whatsoever, to 'evolve' into anything else?

Especially as *with every step away* from the single-celled, bacterial level, *Life gets harder, not easier,* to sustain. WHY would the most successful Life Forms, already established, like Bacteria, need to evolve into anything more complex? Especially as each would step bring *more problems, more challenges* for survival, not less. Darwin's theory is about the *survival* of the *fittest*...and cells, like bacteria, are, *already,* by far and away the *fittest,* with zero need to evolve into anything else.

This is why, far from being science's crowning, Darwin's "Life-is-JUST-CHEMICALS-and-it-AROSE-BY-CHANCE" scheme is not only illogical and unsupported by the data, it is, ironically, *anti-scientific.* Darwin's scheme is unsupported. Its only sort-of-benefit being, for those who dislike all ideas of religion, its claim to get around *'god' or the 'supernatural'* needing to be the ultimate source of anything. It's only logic being, "We can't see Life's *deeper, intelligent, spiritual,* sources standing around in doorways, so they can't be real, can they?

"We can't see *Souls and Spirits* in our telescopes and our microscopes, so they can't exist, can they?"

We can understand some people's frustrations with the traditional religions, and with their officials, who, all too often, preached one thing, then did something different themselves, leading to endless scandals and disappointments for the faithful. This led so many of us, especially after we learned of Darwin's ideas, to throw out the baby of *essential spiritual truths, and basic spirituality,* with the 'bath water' of outdated religious positions which *did* need to be questioned.

This becomes even more of a problem for those of us who have very simplistic ideas of *Spirit-Life-Source-Force-Energy.* If we reject *Spirit* because the only way we can think of *God or Intelligent Creative Beingness* is as a cosmic, Father Christmas type figure, in the sky, with a long beard, on a far off cosmic cloud, making and sustaining everything, *right now, this! amazing! now!, and every! amazing! now!, 24/7!,* until the end of time, then, of course, we will question that.

But, if we conceive of *God-Spirit-Supreme-Intelligent-Beingness* as an amazing *Living, Creating, Intelligence, Force which pervades and* inFORMS *everything,* from *Atoms to Cells to Plants to People to Planets to Stars to Galaxies,* we will have no *need* to go for the totally materialistic and *anti-spiritualist* kinds of explanations which Darwin, Dawkins and so many of their fans have been looking for.

We will have no need to insist that 'science = materialism.' This dreary, reality-shrinking, science-restricting ideology which, ironically, ends up being anti-science, precisely because it does so restrict our inquiries, insisting, quite falsely, that we *'know,'* for established, scientific fact, that *only this* world exists, that we *'know'* that Life arises by pure, stupid, CHANCE, when, in truth, we *'know'* no such thing.

17 Evolution as Religion

In 1981, Colin Patterson, then senior paleontologist at the British Museum of Natural History, provoked an academic storm when he put it to fellow academics that Darwin's theory could no longer be considered knowledge, but it had, like a religion, become a *matter of faith.*

He said that, one day, he "had a sudden realization," and that it struck him that, "I had been working on this [evolution] stuff for twenty years, and there was *not one thing I knew about it.* That was quite a shock, to learn that *one can be so misled for so long."* So, for several weeks, he asked "various people and groups… Can you tell me anything you *know* about evolution, *any one thing that you think is true?"*

He said he "tried that question on the geology staff in the Field Museum of Natural History, and the only answer I got was silence. I tried it on the members of the Evolutionary Morphology Seminar in the University of Chicago ... and all I got there was silence for a long time, and then eventually one person said: 'Yes, I do know one thing. IT OUGHT NOT TO BE TAUGHT in high school.' [205]

"I woke up and I realized that all my life I had been duped into taking evolutionism as REVEALED TRUTH in some way. … Now I think many people in this room would acknowledge that, during the last few years … you've experienced a shift from evolution as knowledge to EVOLUTION AS FAITH. I know that's true of me, and I think it's true of a good many of you in here. So that's my first theme. That *evolution and creationism seem to be showing remarkable parallels.* They are increasingly hard to tell apart.' [206] That was 45 years ago now!

Darwin said *On the Origin of Species* was just an introduction to his ideas, which he would, later, back up with FACTS and REFERENCES. Except, his promised second book never came and, without the needed confirming fossil data ever being found, his ideas became a *faith.*

Modern intelligent design theory, or ID, on the other hand, is *not* the *belief* that this or that religious creation story is *literally* true, but that there is *demonstrably* more to *Living Things* than blind chance can do.

We do not need to know how something was *literally* created—be it a Cell, Fish, Human, TV, Car, or PC—to be able to work out that it was ***created by intelligence,*** not by mere luck. We need to remember too that to *describe* the fossil data, which has been done very well by now, is *not* the same as *truly explaining* it. Our top fossil experts discovering that,

> "This set of animals, with no plausible Darwinian-style ancestors, *suddenly appeared,* did *not* morph into new types of creatures, as Darwin predicted, but *stably* remained the same, before, in time, *dying out,* only to be succeeded by later species appearing in the same mysteriously *abrupt* way, before in their turn, disappearing into extinction, only to be followed by further *Life Waves* of similar or different species . . ."

does not tell us what *caused* those sudden *non-evolutionary* appearances of the species. Our modern knowledge of geology belies bible-based young-earth creationism. But it gives no comfort to the illogical Darwinian notion of a mindless, *yet highly progressive*, evolution, culminating in the *amazing*, yet, biologically, totally unnecessary, abilities of humanity: speech, poetry, art, music, science, mathematics.

Why "biologically unnecessary?" Because *evolution*—Bacteria to Fish to Reptiles to Bears to Whales—is meant to be a totally MINDLESS process. "CHEMICALS became 'alive,' for no reason, and started to eat other CHEMICALS, for *no reason!* Then, eventually, some of the CHEMICALS began to *think, to write poetry* and to discover the laws of the universe. Why? For no reason at all! Because there is *no purpose or intelligence* in Nature! It is all blind, meaningless, accident, isn't it?"

Ah, we need to think! If Darwinian evolution were true, why would "Life" ever need to "progress" beyond some *basic, algal or bacterial stage?* Why would it not simply continue, *mindlessly, "selfish-gene"* reproducing algae or bacteria until the end of time—with no *"natural"* need whatsoever to *"progress" or to "evolve"* beyond such a stage?

We need to stop pretending we now *'know'* how Life began, and evolved from there, when we know no such thing. But, to tell young people that, long ago, many weird and wonderful creatures *suddenly* arrived in the Cambrian Explosion, followed by further mysterious Life Waves, and mass extinctions, including the fascinating story of the intriguing dinosaurs, would be fine, because it's what the fossils show.

To admit that we don't currently know, *scientifically* speaking, what caused the diverse species to arise in the first place would also be true.

To concede that Darwin's theory did ***not*** accord with, ***nor*** did it explain the fossil evidence, ***nor*** the similarities in vertebrate limbs, ***nor*** embryology, ***nor*** homologies, ***nor*** genetics, ***nor*** the DNA CODED complexities of Living things, ***nor*** sex, ***nor*** the intriguing convergences and divergences between marsupial and placental mammals ***nor*** the convergence between the mammalian camera eye and the astonishingly clever octopus's camera eye, ***nor*** the spider's amazing web, ***nor*** the bombardier beetle's extraordinary weaponry, ***nor*** the astonishing caterpillar-to-butterfly metamorphosis, ***nor*** vision, ***nor*** hearing, ***nor*** memory, ***nor*** the incredible biochemistry of blood, ***nor*** the heart, ***nor*** our amazing brains, ***nor*** lungs, ***nor*** *A W A R E N E S S,* ***nor*** feeling, ***nor*** *thinking,* ***nor*** speech, ***nor*** language, would also be correct.

To admit that, currently, there are *no* scientific theories, that truly work, to explain *any* of these remarkable things would also be truthful.

To explain to young people that some scientists are *spiritualists* and, others, are MATERIALISTS, for whom, *all* things are *unintelligently* designed, right up to their own and everyone else's *intelligent minds*—however unlikely that may seem—would also be fair.

Young people can also learn that, while science, at its best, is a sincere search for truth, it can be affected, as Richard Lewontin stated, by people's prior commitments to what they already believe to be true, or what they hope to be true. Those scientists, like Darwin and Crick, Lewontin and Dawkins, who believe that only the MATERIAL

UNIVERSE is real, will put forward purely MATERIALISTIC theories. While counter intuitive to most of humanity, they look at *Living nature,* including, *(are we to believe?),* even at their own children, and their own pets, at their own amazing existences, as if its source is *mindless,* with *no meaning or intelligence or purpose,* and this hypothesis, despite its many—really very obvious—flaws can continue to be explored.

Other scientists, currently a minority, incline to the classical, *mind-before*-MATTER thesis. They follow the ancient intuition that the universe is *multi-dimensional,* that Living things *look* intelligently made, because they really *are intelligently made,* and, they say, it may be possible—using modern, scientific means—to demonstrate this.

Although, frankly, commonsense should be enough. Just looking at a Butterfly, Bear or Whale, Wolf, Eagle or Snail, Infant, Elephant, or Plumber, Physicist, or Dancer should be enough.

While some things in Nature are, clearly, somewhat random, and 'unintelligently designed,' mountain ranges and river valleys, for example, we don't need to be *biologists* to work out that certain of its features, especially in *living* Nature, and *bright* human minds, able to pursue the *arts* and the *sciences*, could never fall together by blind, probability-defying, entropy-violating, *chance, n*o matter how much time we allow, and, no matter how *ideologically annoying* we may find it, be it because we dislike the religions or for whatever other reason.

At the same time, we can make it clear to young people that modern hypotheses of *intelligent design (ID)* and *super* natural causation do not give credence to any particular set of spiritual or religious claims.

However, as evidence supporting the modern ID view, young people can learn of the cell membrane dilemma highlighted by Jacques Monod.[207] He wondered: 'How could the cell's clever, selectively permeable membrane—which protects the DNA inside the cell from being damaged by the outside—arise, absent the very DNA which CODES for it in the first place? They could learn that this, 'chicken or egg?' issue, is an example of *irreducible complexity*—which points, more logically, to *intelligent* causality, than to mere chance.

For those who want *scientific* proof that reality is *intelligent,* not *'dumb,'* we can mention serious modern thinkers, like Behe, Meyer, and Dembski, who use powerful modern techniques for working out whether certain patterns in nature are ra***n***d*O*m or meaningful.

For example, the Search for Extra-Terrestrial Intelligence or SETI is a branch of modern intelligent design detection science, as are archeology, (is that mound a natural feature or is it artificial?), forensics, (was the deceased pushed or did s/he fall?), cryptography (is that gibberish or is it **CODE**?), and that SETI is based on the analysis of radio signals from space for signs of meaningful non-randomness.

In fact, researchers have now realized that if portions of Life's **DNA CODING** were detected, by us, in the form of radio waves coming from space, the conclusion of *intelligence and artificiality* would be unavoidable.[208] Why? Because of the *specific* complexity of Life's **DNA** instructions: **TTAAGGCC + ATAGGAGC + CAACGGTT + CCTTGTAC + TATAGGCC + AGGAATGC.** This is not random or *Shannon* information. In fact, researchers, like shCherbak and Makukov, have "noted various indicators of artificiality in the [DNA] CODE, such that nucleon sums are multiples of 037, that the stop codons act as zero in a decimal system, and all the three-digit decimals (111, 222, 333, 444, 555, 666, 777, 888, and 999) appear at least once in the code, which also looks like an intentional feature."[209] They write that,

"NO NATURAL PROCESS can drive mass distribution to produce the balance ... amino acids and *syntactic signs* that make up this balance are ENTIRELY ABSTRACT since they are produced by translation of a string read across codons. ... not only the signal itself reveals intelligent-like features [but]... taken together all these aspects point at [the] A R T I F I C I A L nature of the patterns."[210]

In these ways, young people can learn that science is not a dogmatic *ontology* or *belief* system, like **MATERIALISM**, but it's about keeping an open mind, and finding out what is *true.* Is Darwin's *materialist* hypothesis of Life more likely *true?* Or the classical *spiritualist* hypothesis of thinkers like Plato and Galileo, Newton and Planck? The fact that we don't currently know, scientifically speaking, exactly *how, or why,* the *stunningly* complex functioning in living things arises, *second, by second, 24/7, our hearts beating, right now!, now!, and now!, our astonishing blood circulating, now!, now, and now!,* nor exactly *how, or why,* the diverse plants, animals, and people arise, does not tell us that there are no intelligent processes involved . . .

Having said this, we can, of course, give young people the facts that we *do* know. "Life on Earth appeared billions of years ago, the Cambrian Explosion took place, trilobites appeared, spiders, snails, and fish, the dinosaurs came and went, birds and bats emerged," and explain to them that there is more than one way to interpret the data.

While making them aware of the *current,* majority view, expressed by thinkers like Darwin, Dawkins and Lewontin, we can tell them that there are increasing numbers of thinkers, like Milton, Denton, Johnson, Behe, Battson, Latham, Dossey, Sheldrake, Radin, Tour, Meyer, Axe, Gauger, Wells, Witt, Dembski, Flannery, Shedinger, Nagel, even some famous *former atheists,* like Anthony Flew,[211] who doubt that our now knowledge of the astonishingly complex workings of Living things is good evidence for 'purposeless' and 'unintelligent' processes in nature.

Why should students not be able to cope with knowing that there are these two basic, yet (very) opposed, philosophies—MATERIALISM *vs Spiritualism*—and scientists like to debate which is correct?

We can explain that there are these two world views, and we can do our best not to mislead them into thinking (a) that physical scientists have the only valid answers, a delusion which is called *scientism,* (b) that it *really was* established, by Darwin, that Life is just an accident, which evolved by accident, nor (c) that 'science' has proved MATERIALISM to be true and *Spiritualism* to be false.

What Colin Patterson highlighted with his question: 'Can you tell me anything you know about evolution *that you think is true?,'* is that Darwin's theories have gone from being once interesting ideas—open to testing and to falsification—to a *faith* position. This is why, despite their repeated falsification, Darwin's ideas are now treated like 'answers' to

Life's mysteries which cannot be questioned, rather than fascinating questions which, as yet, we cannot fully answer.

One odd thing, though. If you ever listen to a discussion about how science is *supposed* to be, you'll be told how scientists always keep their minds open to new ideas and to fresh evidence relating to older ideas, and they are ever ready to revisit their earlier theories, and to humbly follow the data wherever it may lead. In many areas of *practical* science this is true, but, when it comes to Darwin, not so much.

We have seen, over and over again, how the data doesn't match Darwin's predictions. Yet, the data is ignored—over and over again.

We have seen, how methodological materialism and methodological naturalism, MM/MN, while sometimes useful, can, ironically, become science-blockers—when they are used to close down other perfectly reasonable lines of inquiry. We have seen how courageous modern Galileos—like Behe, Meyer and Dembski—are ridiculed or ignored.

Yet, although Darwin's theory has failed to properly explain natural history, let alone *"The [actual] Origin of Species,"* which was it's whole point, relatively few seem interested. It seems odd but it's true.

Yet, our *own* purposeful desires, our own clever *capacities* to do *logical—intelligent—science,* ought to tell us that mindless **CHEMICAL COINCIDENCES,** in no matter how many 'warm little ponds,' cannot possibly explain the arising, in us, of the very faculties which we call *Soul and Mind, Intelligence and Curiosity, Will and Purpose, Experience and Memory, Imagination and Creativity,* all *non-material, Spiritual, faculties* which inspire us to pursue the very activity which we call *logical, rational, Science.*

But, if we adopt *materialism* as our basic creed, then, yes, we *will* be forced to reject all notions of intelligence and purpose in Nature. A creed which takes us away from open minded inquiry and forces us into the *science-confining—inquiry-limiting—ideology* we call *materialism.*

This dreary ideology which makes us, and our science, meaningless. For, if, we can call a modern Jaguar Car or Jaguar Fighter Jet a cleverly designed object of beauty and purpose, but are to pretend that its living, feline inspirations, vastly more clever, more beautiful, and *more complex* than anything we can make, using *all* our intelligence, are, merely, the unintended byproducts of an impossibly *implausible* series of random CHEMICAL COINCIDENCES, which only *seem Soulful, Purposeful, and Alive,* then parts of our sciences are no longer instruments for truth, but misleading tools, for reality-distorting lies, and, whether we realize it or not, they become *mephistophelian.*[212]

But given that, despite all the pro-Darwin claims, there was *never* any good evidence for Darwin's macro theory, and this was pointed out decades ago, why do so many of us still cling to his ideas?

Well, it's partly because, as Chris Carter points out,

> "In our modern world, *science and scientists* hold a great deal of prestige, and few people want to be thought of as unscientific. ... [So] If to be scientific is good and unscientific bad, and if the term "scientific" is thought to be synonymous with the term "materialistic," then any talk of *disembodied minds or spirits* [or *intelligent causation* in Nature] is anti-materialist, unscientific, and therefore bad.[213]
>
> "The long-standing *confusion of materialism* with science is what largely accounts for the persistent social taboo responsible for the ignorance and dismissal of the substantial amount of evidence that proves materialism false." [214]

"But," Carter adds, "the difference between science and ideology is not that they are based on different dogmas; rather, it is that *scientific beliefs* are *not* held as dogmas, but are *open to testing and hence possible rejection."*[215] Bruce Greyson, a prominent near death researcher makes a similar point. He writes,

> "Materialists often claim credit for the scientific advances of the past few centuries. But it is the *scientific method* of empirical *hypothesis testing,* rather than a materialistic philosophy, that has been responsible for the success of science in explaining the [physical] world[s]. If it comes to a choice

> between empirical method and the materialistic world view, the true scientist will choose the former." [216]

Unfortunately, materialism, for many, today, is not "open to testing and possible rejection" and, too often, its dogmas go unquestioned:

> "Only PHYSICAL stuff is real. Blind chance is *the* great creative principle. *Life* is just an accident, which evolved by accident. Therefore, "You, your joys.., your memories and free will, are no more than the behavior of a vast assembly of nerve cells... You're nothing but a pack of neurons."' [217]

This unrealistic ideology which asks us to believe so much non-sense, just to avoid the *classical idea* that we are born of an *intelligent mystery,* which ***just Is,*** rather than a MINDLESS *mystery,* which ***just is.***

For too long have we been presented with a *false* choice: Either we are to believe that the world was instantly created by an absurdly simplistic, Father Christmas type god (false), or, it's far more ancient, (true), but it cannot be a product of *intelligent* causation at all (false). It is all pure accident, all the way up and all the way down. (false).

Patterson was right. Nothing about Darwinian macro-evolution is true. "Darwin said so, so it must be right," is *not evidence. Wikipedia* saying: "most scientists believe that Darwinian evolution is true," is *not evidence.* Imaginative drawings of *imaginary* transitional species, never to be seen or found in the real, *non-imaginary,* fossil records, are *not evidence*. Darwin, in his day, had an idea. It was worth considering.

Yet, too many of us do not notice the glaring incongruence when Darwin's supporters say, "Finch beaks can change," 'Yes, we know that,' "Moths can vary," 'Yes, that's also true,' "and there are many different varieties of pigeon and dog," 'No problem, we all know these things.' But they don't stop there, oh dear me, no. For, they continue:

"So, this proves, as Darwin suggested to his friend Hooker, that DEAD CHEMICALS in a 'little pond,' struck by lightning, could turn into stunningly complex, DNA Coded, *Living,* self-reproducing, *Cells!" 'What?!* That's absurd. It is *chemically and mathematically impossible!,'* call out some who are still paying attention. Ignoring them, our materialist friends continue, "Then, some of these *single* cells morphed into multi-celled worms, spiders and fish, and *sexual reproduction* started, by pure chance."

'Really?,' the still skeptical, try asking. 'Has anyone ***shown*** that this happened? Is there any *evidence* that it did? Let alone *how* it did?'

Taking no notice, Darwin's fans press on: "Then, water breathing fish, after undergoing yet more *random genetic copying error* changes, became *air breathing* amphibians," *'HOW* on earth..?!' Yet, no answer comes back, except: "Then dinosaurs, then birds, with *altogether different* lungs again, then mammals, and, eventually, brilliant minds like Plato, Newton and Einstein." A scheme which is not only unsupported by the data but one which is *demonstrably impossible.*

Isn't it very odd that all this is not seen more critically? Changes over time? Yes, no problem. The mysteriously *abrupt* emergences of new species? Yes. Similarities in design? Yes. The process Darwin imagined? *Sorry, no.* An instant creation, 6000 years ago, based on Bishop Ussher's too literalistic reading of the Bible? *Sorry, also no.*

Right now, all we *really* know is that Life on Earth is very old, and it has gone through vast changes over vast time.

Beyond this, we still know very little, *scientifically* speaking, about Life's true origins, what drove the mysteriously *sudden* appearances of new species, the long periods of *stability,* the eventual *disappearances,* followed, in time, by further, mysterious Life Waves...

Currently, we have no useful ideas as to what causes any *Living Being* to be, or *Mind, or Emotions* to arise out of the apparently MINDLESS DUST of this world, *this! astonishing! second!,* and *every! astonishing second!, 24/7*. No one in mainstream science today has any useful idea how *Cells! or Breathing, Feeding! or Feeling, Sex! and sexual Reproduction, Female! and Male, blood!, and pulsing Hearts!,* all *truly* started or why our physical bodies, eventually, cease working, while, we, as immortal *Souls,* leave this dimension, at least for a while... The MINDLESS ATOMS of the materialist faith are not *alive* in the way we are alive, nor do they need to *breathe!, or feed!, or excrete!, or sleep! or die* as we do. Yet our sciences are, currently, *remarkably incurious* about the true origins of all these fascinating qualities and functions. They have nothing useful to say about them. As they deny the inherent *intelligence!, in the astonishing!, amazing!, living!, Nature,* all around us, as they deny the *Life Forces!, the Chi!, the Prana!, the Soul, and the Spirit!,* in all things, even, in themselves, how can they?

Now, something interesting I've noticed, when it comes to our thinking on Life's origins, is that many of us today are either unfamiliar with the classical notion of *'mystery,'* or, we actively dislike it. We want all things to be worked out in totally linear and MECHANICAL ways.

Some call this approach, *reductionism, or scientism,* the worship of physical science, almost like a religion, in the hope that, eventually, we will bring *all* of reality under our *scientific and technical control.*

Of course, it's natural to want to understand and to control things. Let's try to find out as much as we can about reality, about what is *really true,* including Life's origins. Yet, let's also not forget that, no matter *how* far we go in our researches, all we will manage to do is to penetrate just a little further into the *deepest* mystery of all, which is:

"Why is there any-thing at all, rather than no-thing at all? *Why* is there a *final Source of All,* which ***Just Is?"*** Be it 'dumb,' as our materialist friends argue, or intelligent, as our more *spiritually* minded colleagues believe? No one can say. The *Final Source of All,* be it mindless or be it *Intelligent,* ***just Is***. The *Final Source of All,* be it *soulless and dead, or, be it Intelligent!, and Astonishingly Alive!,* is the *deepest* mystery of all, and, it is not a mystery we will ever solve.

Rational thinking and logical inquiry, wonderful as they are, can never tell us, *'**Why***** there can be any *Being or Thing* rather than none?'

After all, why should there be *any Being-ness or Thing-ness* at all? Even one atom, let alone the ***knowing*** of it? None can say. *Yet there is 'Being,' and 'Isness,' and the Knowing of It.* So, using our logic and our intuitions, we can do our best to make sense, both of its *multitudinous parts,* Atoms, Cells, Plants, Animals, People, Planets, Stars, and of its *unitive totality*—be it a grand theory of everything, as our *scientists* seek, or *a direct mystical experience* of the *Oneness* of everything, as our *mystics and yogis, our shamans, and sufis, like Rumi,* report to us.

Which *"impossible"* of the seemingly *impossible* possibility of there being anything-at-all *vs* nothing-at-all makes more sense? The 'mindless impossible' of MATERIALIST thinking? Or, the 'intelligent impossible' of classical *Spiritualist* thinking?

If only we could just allow ourselves to admit that, currently, we do not have even the whisper of a real idea, *scientifically* speaking, what *Life! and Mind!, Feeling! and Needing!, Love! and Joy!, Curiosity and Compassion* truly are, nor what causes any *Being! or thing* to be so *magically Alive!,* and so *mysteriously Sentient!,* in this, *our beautiful, enchanted, world,* we'd be on much firmer, if humbler, ground.

The current mainstream *science* view, expressed ardently by thinkers like Darwin and Dawkins *'solves'* existence's greatest mystery, the existence of *any.thing* at all *vs.* no.thing at all, by proclaiming:

"Sorry folks, there's nothing *truly* mysterious, *or Intelligent, or Miraculous, or Spiritual* to see in regard to Life, to the very fact of *Existence-Life-Sentiency-Being vs* oblivion and total non-being.

"It is all just MINDLESS, BLIND-CHANCE, chemistry, with no more *'Soul'* or *'Intelligence'* to it than there is to a randomly evolving river valley. There is no more 'meaning' to it all than we, ourselves, care to make up. This is the 'rational' view, the 'scientific' view, the 'logical' view. Now, go forth and enjoy your meaningless existences!"

This is the dreary, inquiry-limiting, science-crimping, reality-shrinking story so many of us now believe in, thanks, in important part, to Darwin and his many (too) loyal followers.

Yet, as to *why,* let alone *how,* the MINDLESS ATOMS of the modern materialist *faith* would ever become Eagle, or Lion, or People shaped, let alone begin to *Think,* and to *Create,* or to soulfully *Care about anything at all, to Love, to Enjoy, and to be Curious, to seek meanings, and to pursue rational, scientific, philosophical, and spiritual inquiries,* Darwin's 'religion' of a blind, random, evolution has little useful to say.

Selected Bibliography — Further Reading

Below are some of the works drawn on in exploring these themes.

On the Origin of Species by Charles Darwin, 1859. Darwin's materialist origins story. Darwin dismissed the lack of convincing fossil evidence for his scheme by appealing to poor fossil preservation. But he conceded that those who did not find this idea convincing, "will rightly reject my whole theory."

Evolution: A Theory in Crisis by Michael Denton, 1986–2016. Updated. Must reading for a more detailed understanding of why Darwin's famous theory has failed to deliver. One of the inspirers of the modern ID movement.

Darwin's Enigma, Luther Sunderland, 1988. Darwin conceded that a real world lack of transitional fossils showing earlier forms morphing into different ones *would falsify his theory.* Sunderland interviews leading fossil authorities who confirm that—in conflict with Darwin's ideas—there is *no fossil evidence* for evolution as he described it. Darwin's theory *was falsified decades ago* but, strangely, most schools, universities, and intellectuals, take no notice.

Shattering the Myths of Darwinism: A rational criticism of evolution theory, Richard Milton, 1992–2017. Darwinism the *scientific* theory has *failed.* It has been replaced by *Darwinism the faith.* The *faith* is said to be "fact" even as it "totters atop a shambles of outdated... evidence" which should have been rejected long ago. R. Dawkins and *Nature* magazine took Milton's critiques seriously enough to decide to launch scathing attacks on his book.

Darwin on Trial, Phillip Johnson, 1993. A bestseller. Also, ***Reason in the Balance: The Case Against Naturalism [Materialism] in Science.***

Darwin's Black Box, The Edge of Evolution, Darwin Devolves, M. Behe, 1998-2019. Explorations of the modern, intelligent design (ID) hypothesis of *irreducible complexity,* and the very limited creative powers of Darwinian evolution. Behe's powerful arguments sink Darwin's highly speculative ship.

Icons of Evolution: Science or Myth? Jonathon Wells, 2002. Examines the surprisingly few, alleged, 'proofs' for Darwinian evolution. He shows that they are all misleading, misrepresentations of evidence, or downright fraudulent.

The Naked Emperor: Darwinism Exposed, Anthony Latham, 2005. Discusses the troubling lack of real world evidence for large scale, long term, evolution as Darwin imagined it. Excellent, readable, concise book.

Billions of Missing Links: A Rational Look at the Mysteries Evolution Can't Explain, Geoffrey Simmons, 2007. Examines some of the many biological functions "that came about all at once, entire…with no preceding links, no subsequent links, *no 'sideways' links. ... Nature contains only LEAPs, not links."* Only *supernatural intelligence* can, truly, explain Life's complexities.

Signature in the Cell: DNA and the Evidence for Intelligent Design, Stephen Meyer, 2009. Codes, including **DNA CODES** *cannot* write themselves. This points to intelligent design, not to blind chance, as materialism urges.

Intelligent Design Uncensored, William Dembski and Jonathan Witt, 2010. Using *modern, scientific* techniques, shows how to determine whether the design in nature—which we can all see—is random or *real* and *meaningful.*

Alfred Russell Wallace: A Rediscovered Life, Michael Flannery, 2011. At first, evolution was referred as 'the Darwin and Wallace theory.' But Wallace, unlike Darwin, realized, in time, that scientific-materialism was inadequate to explain Life and endorsed *spiritualism.*

Darwin's House of Cards: A Journalist's Odyssey Through the Darwin Debates, Tom Bethell, 2016. Having met many influential figures in the evolution world, he concludes that, although Darwin's theory is still touted, it is now *"disintegrating* under an onslaught of new scientific discoveries."

Evolution's Final Days, John Morrison, 2019. Excellent, brief book.

Taking Leave of Darwin, Neil Thomas, 2021. Once a convinced Darwinist, he explains why he now considers Darwin's famous macro theory to be falsified.

Darwin's Bluff: The Mystery of the Book Darwin Never Finished, Robert Shedinger, 2024. Darwin promised he would make good his bluffs in *"On the Origin of Species,"* on the powers of natural selection to evolve all Life, in a later book. Darwin gave up on his *later* book when he realized he could not back up his ideas. He could not find the needed evidence. Darwin's draft MS ***was*** published in the 20th century but added little to *On the Origin of Species.*

Journal of Discovery Institute, ***Evolution News & Science Today:*** exposes how the public continue to be misled into thinking Darwin's ship still floats.

Science and the Near Death Experience, Chris Carter, 2010. How near-death experiences or NDEs provide glimpses of an awaiting *afterlife*. ***Science and the Afterlife Experience: Evidence for the Immortality of Consciousness,*** Carter, 2012. *Mind* and *Soul* exist beyond biology. ***Science and Psychic Phenomena: The Fall of the House of Skeptics,*** Carter, 2012. Evidence that 'telepathy, clairvoyance, precognition, and psychokinesis are all real.

End Notes

[1] It is **materialist-atheism** the belief that only the PHYSICAL WORLDS of atoms like carbon and molecules like H2O, are real, that there is *no Soul or Life after Death,* no *Meaning or Purpose to Existence,* which this work questions. **Idealist-atheism**, which allows for the *Supernatural and the Spiritual* to be the ultimate sources of reality, is not questioned here. Buddhism and Taoism are idealist-atheist traditions. All references to atheism are to materialist-atheism not idealist-atheism.

[2] Darwin to J.D. Hooker. 2.2.1871. Emphasis and text in square brackets added.

[3] Davies, P. 1998. *The Origin of Life.* Penguin.

[4] McLuhan, R. 2010. *Randi's Prize: What Sceptics Say About the Paranormal, Why They Are Wrong, and Why It Matters.* Carter, C. 2012. *Science and Psychic Phenomena: The Fall of the House of Skeptics,* Inner Traditions.

[5] Carter, C. 2010. *Science and the Near Death Experience.* Inner Traditions; p.235.

[6] Gustus, S. 2010. *Less Incomplete: A Guide to Experiencing the Human Condition Beyond the Physical Body.* O Books. Monroe, R.A. 1972. *Journeys Out of the Body.* Souvenir Press. Buhlman, W. 2001. *The Secret of the Soul: Using Out-of-Body Experiences to Understand Our True Nature.* Harper One. Peake, A. 2011. *The Out of Body Experience: The History and Science of Astral Travel.* Watkins

[7] Goswami, A. 1993. *The Self-Aware Universe: How Consciousness Creates the Material World.* Jeremy P Tarcher. Capra, F. 1975.The *Tao of Physics. Shambhala.* Davidson, J. 1992. *Natural Creation or Natural Selection? A Complete New Theory of Evolution.* Element. See also Blavatsky and Steiner.

[8] Darwin to J.D. Hooker. 2.2.1871. Emphasis and text in square brackets added.

[9] Shroeder, G. L. 1996. *Genesis and the Big Bang.* Bantam Doubleday Dell

[10] *Wikipedia,* at time of writing.

[11] Hoyle, F and N.C. 1981. *Wickramasinghe, Evolution from Space. J.M. Dent & Sons.*

[12] Eastman, M & C. Missler. 1995. *The Creator Beyond Time and Space. Emphasis added.*

[13] Denton, M. 1985. *Evolution a Theory in Crisis.* Burnet Books, p.250, emphasis added.

[14] Darwin to J.D. Hooker. 2.2.1871. Emphasis and text in square brackets added.

[15] *Ibid. p 328.*

[16] Schwartz, G.E. Phd. W.L. Simon. 2002. *The Afterlife Experiments: Breakthrough Scientific Evidence of Life After Death.* Atria Books. Fontana, D. 2005. *Is there an Afterlife: A Comprehensive Overview of the Evidence.* O Books.

[17] Richard Lewontin (1997) Billions and billions of demons (review of The Demon-Haunted World: Science as a Candle in the Dark by Carl Sagan, 1997). *The New York Review, January 9, p. 31.*

[18] Lennox, J. 2009.God's Undertaker*: Has Science Buried God?.* Lion. P. 70. Lennox quoting Paul Davies, emphasis and text in square brackets added.

[19] Ross, H. 1995. *The Creator and the Cosmos. Navpress.* p. 117.

[20] Hoyle, F and N.C. 1981. Wickramasinghe, Evolution from Space. J.M. Dent & Sons. Emphasis and text in square brackets added.

[21] McLuhan, R. 2010. *Randi's Prize: What Sceptics Say About the Paranormal, Why They Are Wrong, and Why It Matters.* Carter, C. 2012. *Science and Psychic Phenomena: The Fall of the House of Skeptics,* Inner Traditions.

[22] Gustus, S. 2010. *Less Incomplete: A Guide to Experiencing the Human Condition Beyond the Physical Body.* O Books. Monroe, R.A. 1972. *Journeys Out of the Body.* Souvenir Press. Buhlman, W. 2001. *The Secret of the Soul: Using Out-of-Body Experiences to Understand Our True Nature.* Harper One. Peake, A. 2011. *The Out of Body Experience: The History and Science of Astral Travel.* Watkins

[23] Carter, C. 2010. *Science and the Near Death Experience.* Inner Traditions.

[24] Rupert Sheldrake's Website: Dialogues and Controversies; emphasis and text in brackets added.

[25] Schwartz, G.E. Phd. W.L. Simon. 2002. *The Afterlife Experiments: Breakthrough Scientific Evidence of Life After Death.* Atria Books. Fontana, D. 2005. *Is there an Afterlife: A Comprehensive Overview of the Evidence.* O Books.

[26] Stevenson, I. 1980. *Twenty Cases Suggestive of Reincarnation: Second Edition, Revised and Enlarged.* University of Virginia Press. Stemman, R. 2012. *The Big Book of Reincarnation: Examining the Evidence that We Have All Lived Before.* Hierophant Publishing.

[27] Gates, B. 1996. *The Road Ahead.* Penguin. p228.

[28] It is **materialist-atheism** the belief that only the PHYSICAL WORLDS of atoms like carbon and molecules like H2O, are real, that there is *no Soul or Life after Death,* no *Meaning or Purpose to Existence,* which this work questions. **Idealist-atheism**, which allows for the *Supernatural and the Spiritual* to be the ultimate sources of reality, is not questioned here. Buddhism and Taoism are idealist-atheist traditions. All references to atheism are to materialist-atheism not idealist-atheism.

[29] Dennett D. 1992. *Consciousness Explained.* Back Bay Books, p. 406.

[30] Crick F. 1995. *The Astonishing Hypothesis.* Scribner, p. 3, text in square brackets added.

[31] Dawkins, R. *River Out of Eden: A Darwinian View of Life.*

[32] Carter, C. 2010. *Science and the Near Death Experience.* Inner Traditions, amazon, T. Skaftnesmo, text in square brackets added.

[33] Ibid. p. 261, emphasis added.

[34] 'In practice, laboratory experiments have shown that right-handed building blocks have a slightly greater affinity, or attraction, for other right-handed building blocks. Therefore, at each step in the addition of another building block, there is a 3/7 chance that the next one added will be the same optical isomer as the one previously added. . . . The smallest known free living life forms, bacteria, have about 12,000,000 nucleotides in their DNA. If we were to calculate the odds of adding twelve million successive right-handed nucleotides to the growing chain, without a single left- handed one being added, it would be 5 raised to the 12 millionth power, (5/12,000,000)!' Eastman & Missler

[35] Eastman, M & C. Missler. 1995. *The Creator Beyond Time and Space.* Emphasis, alternative fonts, and text in square brackets added.

[36] Orgel, L. 1973.*The Origins of Life. John Wiley;* p. 189. Emphasis added. Learnt of Orgel's work in S. Meyer's *Signature in the Cell.*

Flew, A. 2009. *There is a God: How the World's Most Notorious Atheist Changed His Mind* Harper One.

[38] The Hottest New Computer Is: DNA *www.evolutionnews.org/2017/10/03*

[39] Richard Dawkins. 2008. *Life: A Gene-Centric View, Craig Venter & Richard Dawkins: A Conversation in Munich. The Edge.* Emphasis added.

40 Michael Polanyi, "Life's Irreducible Structure," *Science 160 (1968), 1308–12,* quoted by Wells, J. 2006. *Darwinism and Intelligent Design.* Regnery Publishing; pp. 100–101, emphasis added.

See *www.evolutionnews.org*/2017/03/hottest-new-computer-dna

42 Denton, M. 1985. *Evolution: A Theory in Crisis.* Burnett Books, p. 260, emphasis added.

Denton, M. 1985. *Evolution: A Theory in Crisis.* Burnett Books.

44 Monod, J. 1972. *Chance and Necessity.* Collins. pp 134–135, text in square brackets added.

45 Latham, A. *The Naked Emperor: Darwinism Exposed.* London: Janus, p. 13. Emphasis added.

46 uncommondescent.com/intelligent-design/a-world-famous-chemist-tells-the-truth-theres-no-scientist-alive-today-who-understands-macroevolution.

47 uncommondescent.com/intelligent-design/a-world-famous-chemist-tells-the-truth-theres-no-scientist-alive-today-who-understands-macroevolution/#:~:text=Although most scientists leave few,to me, permit unhealthy leeway

48 Quoted from George Wald, 'The Origin of Life', *Scientific American* 191:48 (May 1954). Emphasis and text in square brackets added.

49 Quoted from George Wald, 'The Origin of Life', *Scientific American 191:48 (May 1954).*

50 Latham, A. 2005. *The Naked Emperor: Darwinism Exposed.* London: Janus, p. 39.

51 Gould, S.J. *Wonderful Life. 1990. The Burgess Shale and the Nature of History,* New York: Vintage, pp. 59–66; emphasis and text in square brackets added.

52 *Ibid.*

53 Denton, M. 1985. *Evolution: A Theory in Crisis.* Burnett Books.

54 Darwin, C. 1996. *On the Origin of Species.* Oxford University Press, p. 249.

55 Meyer, S. C. 2013. *Darwin's Doubt: The Explosive Origin of Animal Life and the Case for Intelligent Design.* Harper One. Book description.

56 *Darwin's Doubt,* amazon review, C. Luskin. Emphasis added.

57 Darwin, C. 1859. *On the Origin of Species by Means of Natural Selection or the Preservation of Favoured Races in the Struggle for Life,* reprint of 6th edition (John Murray, 1902), p. 341–342. Emphasis added.

58 Goldschmidt, R.B. 1940. *The Material Basis of Evolution.* Yale University Press; p. 390.

59 Sunderland, L. 1988. *Darwin's Enigma: Ebbing the Tide of Naturalism.* Emphasis added.

60 Battson, A. *On the Origin of Stasis by Means of Natural Processes.* Battson's website. Emphasis and text in square brackets added.

61 Lennox, J. 2009. *God's Undertaker.* Lion; p. 143, emphasis added.

62 *Ibid.*

63 Milton, R. 1992. *Shattering the Myths of Darwinism,* Inner Traditions, emphasis, text in square brackets added.

[64] Latham, A. *The Naked Emperor: Darwinism Exposed.* London: Janus, p. 61 emphasis and words in square brackets added.

[65] Battson, A. 1997. *Facts, Fossils, and Philosophy.* Author's website. Emphasis and text in square brackets added.

[66] Darwin, C. 1859. *On The Origin of Species. Ch.VI.*

[67] Darwin, C. 1859. *On the Origin of Species by Means of Natural Selection or the Preservation of Favoured Races in the Struggle for Life.* Emphasis and capitalizations added.

[68] Latham, A. *The Naked Emperor: Darwinism Exposed.* London: Janus, p. 76, emphasis and words in square brackets added.

[69] Ibid, p. 77 emphasis and words in square brackets added.

[70] *https://evolutionnews.org/2018/08/inexplicable-species-and-the-theory-of-evolution/*

[71] Ibid, p. 77 emphasis and words in square brackets added.

[72] Morrison, J. 2019. *Evolution's Final Days: The Mounting Evidence Disproving The Theory of Evolution (problems, myth, hoax, fraud, flaws)* Emphasis added.

[73] Gould, S. J. 1985. *"Not Necessarily a Wing" Natural History, October, pp. 12, 13.* Text in square brackets added.

[74] Lovtrup, S. 1987. *Darwinism: The Refutation of a Myth.* Croom Helm Ltd., Beckingham, Kent, p. 275, text in square brackets added.

[75] Michael Flannery. 'Was Darwin a Scholar or a Pitchman?' *Evolution News,* October 20, 2015. Emphasis and text in square brackets added.

[76] C. Darwin, *On The Origin Of Species*, Emphasis, text in square brackets added.

[77] Davidson, J. 1992. *Natural Creation or Natural Selection – A Complete New Theory of Evolution.* Element; p. 13.

[78] Denton, M. 2002. *Nature's Destiny: How the Laws of Biology Reveal Purpose in the Universe.* Free Press. p. 342

[79] C. Darwin, *The Origin Of Species*, Chapter X, "On the Imperfection of the Geological Record. Emphasis added.

[80] Bethell, T. 2016. *Darwin's House of Cards.* Discovery Institute, emphasis and text in square brackets added.

[81] Demolishing Darwin's Tree: Eric Bapteste and the Network of Life. *Evolution News.* 9.9.13. Emphasis and text in square brackets added.

[82] Johnson, P. 2010. *Darwin on Trial.* IVP Books, p. 17.

[83] Gould, S. J. 1980. *The Panda's Thumb.* p. 181-182)

[84] Simpson, G.G. 1944. *Tempo and Mode in Evolution.* Columbia University Press; pp. 105, 107. Emphasis and text in square brackets added.

[85] Sunderland, L. 1988. *Darwin's Enigma: Ebbing the Tide of Naturalism.* Emphasis added.

[86] Sunderland, L. 1988. *Darwin's Enigma: Ebbing the Tide of Naturalism.* Emphasis added.

[87] Ibid

[88] Stevenson, I. 1980. *Twenty Cases Suggestive of Reincarnation: Second Edition, Revised and Enlarged.* University of Virginia Press. Stemman, R. 2012. *The Big Book of Reincarnation: Examining the Evidence that We Have All Lived Before.* Hierophant Publishing.

[89] Schwartz, G.E. Phd. W.L. Simon. 2002. *The Afterlife Experiments: Breakthrough Scientific Evidence of Life After Death.* Atria Books. Fontana, D. 2005. *Is there an Afterlife: A Comprehensive Overview of the Evidence.* O Books.

[90] McLuhan, R. 2010. *Randi's Prize: What Sceptics Say About the Paranormal, Why They Are Wrong, and Why It Matters.* Carter, C. 2012. *Science and Psychic Phenomena: The Fall of the House of Skeptics,* Inner Traditions.

[91] Taylor, S. 2010. *Waking From Sleep.* Hay House, p. 44. '. . . to indigenous peoples there are *no such things as "inanimate objects"*. . . . they sensed that even rocks, the soil and rivers had a *being or consciousness* of their own. Whereas European or Asian peoples tend to see these things as nothing more than one-dimensional objects which they are entitled to use for their own devices, *indigenous peoples* felt that the natural phenomena around them were ***alive*** and so treated them with respect.. . .

As anthropologist Robert Lawler notes, "Every distin-guishable energy, form or substance has both an *objective* and a *subjective* [*Spiritual*] expression." Aborigines have complained that the problem with European Australians is that they can only perceive the surface reality of the world and can't enter its *interior* life. As one aboriginal elder stated,

"Unless white man learns to enter the dreaming of the countryside, the plants, and animals before he uses or eats them, he will become sick and insane and destroy himself."'

[92] Davidson, J. 1992. *Natural Creation or Natural Selection? – A Complete New theory of Evolution.* Element

[93] Carter: "For the materialist, the term *unscientific* seems to be the modern equivalent of the term heretical, and it is [used] to exclude from consideration ideas that challenge the believer's faith."

[94] Carter, C. 2010. *Science and the Near Death Experience.* Inner Traditions; p.238. Text in square brackets added.

[95] Carter, C. 2010. *Science and the Near Death Experience.* Inner Traditions; p.238. Text in square brackets added.

[96] It is **materialist-atheism** the belief that only the PHYSICAL WORLDS of atoms like carbon and molecules like H2O, are real, the belief that there is *no Soul or Life after Death,* nor any *Meaning or Purpose to Existence,* which this work questions. **Idealist-atheism**, which allows for the *Supernatural or the Spiritual* to be the ultimate sources of reality, is not questioned here. Buddhism and Taoism are **idealist-atheist** traditions. Any references to atheism are to materialist-atheism not to idealist-atheism.

[97] Article by J. Wells. *https://evolutionnews.org/2008/04/is_the_science_of_richard_dawk/*

[98] Johnson, P. 2010. *Darwin on Trial.* IVP Books, p. 16, emphasis and text in square brackets added

[99] Letter to J.F.W. Herschel, 23.5.1861. www.darwinproject.co.uk

[100] McLuhan, R. 2010. *Randi's Prize: What Sceptics Say About the Paranormal, Why They Are Wrong, and Why It Matters.* Carter, C. 2012. *Science and Psychic Phenomena: The Fall of the House of Skeptics,* Inner Traditions.

[101] Carter, C. 2010. *Science and the Near Death Experience.* Inner Traditions.

[102] Gitt, W. 2006. *In the Beginning Was Information: A Scientist Explains the Incredible Design in Nature.* Master Books, p. 124; emphasis and text in square brackets added.

[103] Lennox, J. *2009. God's Undertaker: Has Science Buried God?* Lion; p. 182, emphasis and text in square brackets added.

[104] Dear Reader, please glance at your book's cover. There is no evidence for single cells evolving—randomly or due to intelligent inputs—into jelly fish or worms, worms into fish, bears into whales, shrews into bats. Just the tired, old assertions that: 'It must be so… because Darwin thought so.'

[105] Lennox, J. 2009. *God's Undertaker.* Lion; p. 143.

[106] Lennox, J. 2009. *God's Undertaker.* Lion; p. 143, emphasis added.

[107] Behe, M. 2008. *The Edge of Evolution.* Free Press

[108] Durston, K. 9.7.15 An Essential Prediction of Darwinian Theory Is Falsified by Information Degradation *Evolution News.* Emphasis added.

[109] Behe, M. 2019. *Darwin Devolves.* Harper One.

[110] Perloff, J. 1999, *Tornado in a Junk Yard* https://jamesperloff.com/why-the-creation-evolution-debate-matters/

[111] Denton, M. 1985. *Evolution: A Theory in Crisis.* Burnett Books. p. 209. Emphasis and text in square brackets added.

[112] Ibid. p. 210–212. Emphasis and text in square brackets added.

[113] Perloff, J. 1999, *Tornado in a Junk Yard* https://jamesperloff.com/why-the-creation-evolution-debate-matters/

[114] Ibid, emphasis, including exclamation marks and text in square brackets added.

[115] Morrison, J. 2019. *Evolution's Final Days: The Mounting Evidence Disproving The Theory of Evolution (problems, myth, hoax, fraud, flaws)* Emphasis added.

[116] Shedinger, R. 2024. *Darwin's Bluff: The Mystery of the Book Darwin Never Finished.* Discovery Institute. Review by Gunter Bechly, on Amazon.

[117] Quoted from George Wald, 'The Origin of Life', *Scientific American* 191:48 (May 1954). Emphasis and text in square brackets added.

[118] Quoted from George Wald, 'The Origin of Life', *Scientific American 191:48 (May 1954).*

[119] Richard Lewontin (1997) Billions and billions of demons (review of The Demon-Haunted World: Science as a Candle in the Dark by Carl Sagan, 1997). *The New York Review, January 9, p. 31.* Emphasis and text in square brackets added.

[120] Michael Flannery. 'Was Darwin a Scholar or a Pitchman?' *Evolution News (October 20, 2015).* Emphasis added.

[121] *Bethell, T. 2016. Darwin's House of Cards. Discovery Institute.*

[122] Quoted by Lennox, J. 2009. *God's Undertaker.* Lion; p. 143, emphasis added.

[123] Latham, A. *The Naked Emperor: Darwinism Exposed.* Janus, p. 88. Text in square brackets added.

[124] Wells, J. 2006. *The Politically Incorrect Guide to Darwinism and Intelligent Design.* Regnery.

[125] Gee, H. 1999. *In Search of Deep Time: Beyond the Fossil Record to a New History of Life.* The Free Press; pp. 32, 113–17.

[126] Davidson, J. 1992. *Natural Creation or Natural Selection – A Complete New Theory of Evolution.* Element; p. 13–14, emphasis and text in square brackets added.

127 Wells, J. 2000. *Icons of Evolution.* Regnery, p. 116.

128 Ibid.

129 Bethell, T. 2016. *Darwin's House of Cards.* Discovery Institute, emphasis and text in square brackets added.

130 C. Darwin, *The Origin Of Species*, Chapter X, "On the Imperfection of the Geological Record. Emphasis added.

131 Denton, M. 1985. *Evolution: A Theory in Crisis.* Burnett Books. Text in square brackets added.

132 Darwin, C. 1872. *On the Origin of Species,* 6th ed, 1962, New York: Collier Books, p. 41.

133 Denton, M. 1985. *Evolution: A Theory in Crisis.* Burnett Books.

134 Johnson, P. 2010. *Darwin on Trial.* IVP Books, p. 60, emphasis and text in square brackets added.

135 www.catholic.com/magazine/online-edition/does-it-matter-that-many-scientists-are-atheists#

136 Eastman, M & C. Missler. 1995. *The Creator Beyond Time and Space. Emphasis added.*

137 Dembski, W. 2007. *No Free Lunch: Why Specified Complexity Cannot Be Purchased without Intelligence.* Rowman Littlefield. 'Very Insightful,' 4. 22.02, amazon review, text in brackets added.

138 Simmons, G. 2007. *Billions of Missing Links.* Harvest House; pps. 207–211. Emphasis added.

139 Ibid, p. 29. emphasis and text in square brackets added.

140 Ibid, pps. 207–211,

141 Geng Tian and Daniel Finley, "Cell biology: Destruction deconstructed," Nature 482, (09 February 2012), pp. 170–171, emphasis added.

142 Evolution News, 16th February, 2012; emphasis and text in square brackets added.

143 Emphasis and text in square brackets added. References cited in the Discovery article: (1) Geng Tian and Daniel Finley, 'Cell biology: Destruction deconstructed', Nature 482, 9 February 2012, pp. 170–171. (2) Lander et al., 'Complete subunit architecture of the proteasome regulatory particle', Nature 482, (9 February 2012), pp. 186–191.

144 Behe, M. 1996. *Darwin's Black Box.* The Free Press, p. 185. Emphasis and text in brackets added.

145 Schwartz, G.E. Phd. W.L. Simon. 2002. *The Afterlife Experiments: Breakthrough Scientific Evidence of Life After Death.* Atria Books. Fontana, D. 2005. *Is there an Afterlife: A Comprehensive Overview of the Evidence.* O Books.

146 Quoted by Michael E. Tymn, http://whitecrowbooks.com/michaeltymn/biography/ "It wasn't long after the birth of modern *Spiritualism* in 1848 that scientists and scholars began investigating the phenomena. Many of them started out with the intent of showing that all mediums were charlatans, but one by one they came to believe in the reality of mediumship and related psychic phenomena." Tymn, Wallace's1874 defense of the spiritualist data, can be found in *Miracles and Modern Spiritualism.*

147 Schwartz, G.E. Phd. W.L. Simon. 2002. *The Afterlife Experiments: Breakthrough Scientific Evidence of Life After Death.* Atria Books.

148 Flannery, M. A. 2011. *Alfred Russell Wallace: A Rediscovered Life,* Discovery Institute Press

149 Since the artificial synthesis of urea in 1828 the science view is that there are no Life Forces.

150 According to science today, all things in Nature are 'coincidences,' and 'purposeless.'

[151] Eastman, M & C. Missler. 1995. *The Creator Beyond Time and Space. Emphasis added.*

[152] George Wald, 'The Origin of Life', *Scientific American* 191:48 (May 1954). Eastman, M. C. Missler. 1995. *The Creator Beyond Time and Space*

[153] Kelly, E. F. E.W. Kelly, A. Crabtree, A. Gauld 2009 *Irreducible Mind: Toward a Psychology for the 21st Century.*

[154] Behe, M. 1996. *Darwin's Black Box. The Free Press.*

[155] Ibid.

[156] Demolishing Darwin's Tree: Eric Bapteste and the Network of Life. *Evolution News. 9.9.13.*

[157] Gould, S. J. 'Is a new and general theory of evolution emerging?' *Paleobiology, vol. 6(1), January 1980,* p. 127. Emphasis added.

[158] Eldredge, N. & I. Tattersall. 1982. *The Myths of Human Evolution.* Columbia University Press, p. 48.

[159] Simpson, G. G. 1955. *The Major Features of Evolution.* Columbia University Press, p.360.

[160] *https://evolutionnews.org/2021/11/*the-discontinuous-fossil-record-refutes-darwinian-gradualism/

[161] Battson, A. Conflicts Between Darwin and Paleontology. *http://www.veritas- ucsb.org/library/battson/stasis/2.html*

[162] Gould, S. J. 1980, emphasis added.

[163] C. Darwin, *On The Origin Of Species*, Emphasis, capitals, bold etc added.

[164] Gould, S. J. 1980. *Is a New and General Theory of evolution emerging?* p.120, emphasis added.

[165] Ibid. p. 61 emphasis and words in square brackets added.

[166] Patterson, C. 1978. *Evolution.* The British Museum of Master Books, Natural History.

[167] Sunderland, L. 1988. *Darwin's Enigma.* p. 89. Emphasis and text in square brackets added.

[168] Milton, R. 2000. *Shattering the Myths of Darwinism.* Inner Traditions Bear and Company; back cover, text in brackets added.

[169] Denton, M. 1985. *Evolution: A Theory in Crisis.* Burnett Books. p. 213. Emphasis and text in square brackets added.

[170] Simmons, G. 2007. *Billions of Missing Links. Harvest House;* pps. 75–76, emphasis and text in square brackets added.

[171] Darwin, C. 1859 *On the Origin Of Species* Ch. X, "On the Imperfection of the Geological Record".

[172] Ibid., Ch. IX, emphasis added.

[173] Sunderland, L. 1988. *Darwin's Enigma: Ebbing the Tide of Naturalism*. Emphasis added.

[174] Darwin, C. 1859. *On the Origin of Species by Means of Natural Selection or the Preservation of Favored Races in the Struggle for Life* of 6th edition (John Murray, 1902), p. 341–342. Emph. added.

[175] Eldredge, N. & I. Tattersall. 1982. *The Myths of Human Evolution.* Columbia University Press, p. 48.

[176] Simpson, G. G. 1955. *The Major Features of Evolution.* Columbia University Press, p.360.

[177] Sunderland, L. 1988. *Darwin's Enigma.* p. 89. Emphasis and text in square brackets added.

[178] Milton, R. 2000. *Shattering the Myths of Darwinism.* Inner Traditions Bear and Company; back cover, text in brackets added.

[179] *Discover Magazine,* p. 68 (April, 2011).

[180] Margulis, L. & D. Sagan. 2003. *Acquiring Genomes: A Theory of the Origins of the Species.* Basic Books, p. 29. Emphasis and text in square brackets added.

[181] Quoted in "Discover Interview: Lynn Margulis Says She's Not Controversial, She's Right," *Discover Magazine, p. 68 (April, 2011).* Emphasis and text in square brackets added.

[182] Mayr, E. 1942. *Systematics and the Origin of Species.* Dover Publications, p. 296.

[183] Malcolm Muggeridge, *The Advocate,* March 8, 1984, p. 17.

[184] Schwartz, G.E. Phd. W.L. Simon. 2002. *The Afterlife Experiments: Breakthrough Scientific Evidence of Life After Death.* Atria Books. Fontana, D. 2005. *Is there an Afterlife: A Comprehensive Overview of the Evidence.* O Books.

[185] Richard Lewontin (1997) Billions and billions of demons (review of The Demon-Haunted World: Science as a Candle in the Dark by Carl Sagan, 1997). *The New York Review, January 9, p. 31.*

[186] Denton, M. 1985. *Evolution: A Theory in Crisis.* Burnett Books.

[187] Demolishing Darwin's Tree: Eric Bapteste and the Network of Life. *Evolution News.* 9.9.13.

[188] Sunderland, L. 1988. *Darwin's Enigma.* p. 89. Emphasis and text in square brackets added.

[189] Wells, J. 2000. *Icons of Evolution.* Regnery. p. 73.

[190] Ibid. p. 74.

[191] Ibid. p. 71.

[192] Ibid. p. 83.

[193] Ibid. p. 101.

[194] Tim White, quoted in Ann Gibbons, "In Search of the First Hominids," *Science (Feb. 15, 2002), 295:1214–1219.*

[195] More detailed discussion of the fossil evidence and human origins, Casey Luskin, "Human Origins and the Fossil Record" in *Science and Human Origins* Discovery Institute Press, 2012, pp. 45–83.

[196] Leslie Aiello, quoted in Richard Leakey and Roger Lewin, Origins Reconsidered: In Search of What Makes Us Human (Anchor Books, 1993), p. 196.

[197] John Hawks et al., "Population Bottlenecks and Pleistocene Human Evolution," *Journal of Molecular Biology and Evolution (2000), 17(1):2–22.*

[198] Ernst Mayr, What Makes Biology Unique? (Cambridge Univ. Press, 2004), p. 198. Emphasis added.

[199] "New study suggests big bang theory of human evolution," (Jan. 10, 2000) at http://www.umich.edu/~newsinfo/Releases/2000/Jan00/r011000b.html. Emphasis added.

[200] Has Science Shown That We Evolved from Ape-like Creatures? Casey Luskin *Salvo magazine September 26, 2013.* Emphasis added.

[201] Language Is a Rock Against Which Evolutionary Theory Wrecks Itself. *Evolution News,* Michael Egnor. 09.19.16. Emphasis and text in brackets added.

[202] Mayr, E. 1942. *Systematics and the Origin of Species. Dover Publications,* p. 296. Text in square brackets added.

[203] Wells, J. 2000. *Icons of Evolution.* Regnery. pp 229 – 230; emphasis and text in brackets added.

[204] Wells, J. 2000. *Icons of Evolution.* Regnery. pp 229 – 230; emphasis and text in brackets added.

[205] *Can You Tell Me Anything About Evolution?* Presentation by Colin Patterson at the American Museum of Natural History. Nov. 1981. Emphasis added. A CD and annotated transcription of the talk, including Patterson's interactions with luminaries like Niles Eldridge in the Q&A and lively discussion that followed the talk was, at one time, available at www.arn.org.

[206] Ibid. Emphasis added. First saw Colin Patterson's famous comments on 'evolution as faith' rather than genuine scientific knowledge, in Phillip Johnson's *Darwin on Trial.*

[207] Monod, J. 1972. *Chance and Necessity.* Collins. pp 134–135.

[208] The 'Wow! signal' of the terrestrial genetic code. *j.icarus.* 2013.02.17

[209] Ibid.

[210] *Evolution News* review of the "Wow!" Signal of Intelligent Design published in the Planetary Science Journal Icarus. *www.evolutionnews.org.* Text in square brackets and emphasis added.

Flew, A. 2009. *There is a God: How the World's Most Notorious Atheist Changed His Mind* Harper One.

[212] Goethe's Mephistopheles tricked Faust into selling his own Soul. The word "could derive from the Hebrew mephitz, meaning "destroyer", and tophel, meaning "liar"... The name can also be a combination of three Greek words: "me" as a negation, "phos" meaning light, and "philis" meaning loving, making it mean "not-light-loving..."' *wikipedia.*

[213] Carter: "For the materialist, the term *unscientific* seems to be the modern equivalent of the term heretical, and it is [used] to exclude from consideration ideas that challenge the believer's faith."

[214] Carter, C. 2010. *Science and the Near Death Experience.* Inner Traditions; p.238. Text in square brackets added.

[215] Ibid, p.237.

[216] Greyson, B.: "Commentary on 'Psychophysiological and Cultural correlates Undermining a Survivalist Interpretation of Near-Death Experiences.'" *Journal of Near Death Studies* 26, no. 2 (Winter 2007); P.142.

[217] Crick F. 1995. *The Astonishing Hypothesis.* Scribner, p. 3, text in square brackets added.

www.ingramcontent.com/pod-product-compliance
Ingram Content Group UK Ltd.
Pitfield, Milton Keynes, MK11 3LW, UK
UKHW020144290926
15471UKWH00004B/689

9 781789 636475